AF324103

Studies
in the History of Mathematics and
Physical Sciences

9

Editor

G. J. Toomer

Studies in the History of Mathematics
and Physical Sciences

Volume 1
A History of Ancient Mathematical Astronomy
By **O. Neugebauer**
ISBN 0-387-**06995**-X

Volume 2
A History of Numerical Analysis from the 16th through the 19th Century
By **H. H. Goldstine**
ISBN 0-387-**90277**-5

Volume 3
I. J. Bienaymé: Statistical Theory Anticipated
By **C. C. Heyde** and **E. Seneta**
ISBN 0-387-**90261**-9

Volume 4
The Tragicomical History of Thermodynamics, 1822-1854
By **C. Truesdell**
ISBN 0-387-**90403**-4

Volume 5
A History of the Calculus of Variations from the 17th through the 19th Century
By **H. H. Goldstine**
ISBN 0-387-**90521**-9

Volume 6
The Evolution of Dynamics: Vibration Theory from 1687 to 1742
By **J. Cannon** and **S. Dostrovsky**
ISBN 0-387-**90626**-6

Volume 7
The Prehistory of the Theory of Distributions
By **J. Lützen**
ISBN 0-387-**90647**-9

Volume 8
Zermelo's Axiom of Choice: Its Origins, Development, and Influence
By **G. H. Moore**
ISBN 0-387-**90670**-3

Bruce Chandler
Wilhelm Magnus

The History of Combinatorial Group Theory:
A Case Study in the History of Ideas

Springer-Verlag
New York   Heidelberg   Berlin

Bruce Chandler
The College of Staten Island of
The City University of New York
715 Ocean Terrace
Staten Island, NY 10301
U.S.A.

Wilhelm Magnus
11 Lomond Place
New Rochelle, NY 10804
U.S.A.

AMS Subject Classifications: 01A60, 20-03, 20BXX

With one illustration .

Library of Congress Cataloging in Publication Data

Chandler, Bruce, 1931-
    The history of combinatorial group theory.
    (Studies in the history of mathematics and physical
sciences; 9)
    Bibliography: p.
    Includes indexes.
    1. Combinatorial group theory—History. I. Magnus,
Wilhelm, 1907- . II. Title. III. Series.
QA171.C447   1982   512′.22   82-10471

Typeset by Science Typographers, Medford, New York.
Printed and bound by R. R. Donnelley & Sons, Harrisonburg, Virginia.
Printed in the United States of America.

9 8 7 6 5 4 3 2 1

ISBN 0-387-90749-1   Springer-Verlag New York Heidelberg Berlin
ISBN 3-540-90749-1   Springer-Verlag Berlin Heidelberg New York

Preface

One of the pervasive phenomena in the history of science is the development of independent disciplines from the solution or attempted solutions of problems in other areas of science. In the Twentieth Century, the creation of specialties within the sciences has accelerated to the point where a large number of scientists in any major branch of science cannot understand the work of a colleague in another subdiscipline of his own science. Despite this fragmentation, the development of techniques or solutions of problems in one area very often contribute fundamentally to solutions of problems in a seemingly unrelated field. Therefore, an examination of this phenomenon of the formation of independent disciplines within the sciences would contribute to the understanding of their evolution in modern times.

We believe that in this context the history of combinatorial group theory in the late Nineteenth Century and the Twentieth Century can be used effectively as a case study. It is a reasonably well-defined independent specialty, and yet it is closely related to other mathematical disciplines. The fact that combinatorial group theory has, so far, not been influenced by the practical needs of science and technology makes it possible for us to use combinatorial group theory to exhibit the role of the intellectual aspects of the development of mathematics in a clearcut manner. There are other features of combinatorial group theory which appear to make it a reasonable choice as the object of a historical study. It is a rather young discipline, being approximately a century old. The literature, although not small (comprising about 5000 papers) was surveyed in 1939 by WILHELM MAGNUS and in 1974 by GILBERT BAUMSLAG. Nearly the entire body of research in the field is due to mathematicians who either are still alive or who were the teachers or senior colleagues of living mathematicians. This makes it possible to supplement the written tradition with oral information which is particularly valuable when dealing with questions of motivation for a particular investigation or of the transfer of ideas.

We have supplemented the mathematical discussions with some bio-graphical data and with general descriptions of the external conditions for mathematical research, using examples from our special field as illustra-tions. In Chapter II.14 we also try to describe some of the effects of the rapid growth of mathematical research.

We gratefully acknowledge the help which we received from many sources in writing this book. We cannot enumerate the names of the many individuals who helped us with information and advice. Apart from these, we wish to thank the National Science Foundation for its generous support which covered all of Part I and without which we would not have begun our project. The hospitality of the Mathematics Research center of Warwick University at Coventry, England enabled us to interview a great many group theorists who visited there in 1978. The Polytechnic Institute of New York, through its Department of Mathematics, provided us with much of the technical and bibliographic help which we needed. And the excellent and well-organized library of the Courant Institute of Mathematical Scien-ces at New York University reduced to a pleasant minimum the normally onerous task of getting hold of the documentary sources for a historical study.

November 1982   BRUCE CHANDLER
                                                 WILHELM MAGNUS

Table of Contents

Part I
The Beginning of Combinatorial Group Theory

Chapter I.1

Introduction to Part I

Combinatorial group theory may be characterized as the theory of groups which are given by generators and defining relations or, as we would say today, by a presentation. Of course, this is not a complete definition of the field and we shall not even try to give one. But at least for Part I it is a fully adequate description of the subject we plan to deal with.

The first problem facing the historian is that of finding a starting point. In our case, this was rather easy. The paper by WALTHER VON DYCK [1882] is the first paper in which generators and defining relations are not only introduced as new concepts, but are also used effectively for mathematical research. It is most likely that at least the germs of the ideas introduced by DYCK can be traced to earlier authors. A thorough study of the emergence of the various aspects of the concept of a group was conducted by H. WUSSING [1969].

Starting with a report on DYCK's first paper and an analysis of its contents, our book describes the unfolding of combinatorial group theory, its concepts, problems, results, and its relation with other fields, in particular, with topology. Here we encounter the second difficulty facing every historian. It is impossible to write a universal history. In the case of mathematics, it is even impossible to explain the technical terms appearing in the influencing disciplines in any but the most superficial manner. We have met the difficulties arising here with a compromise, i.e., something which, by its very nature, cannot be entirely satisfactory.

Part I of our book covers the period from 1882 to 1918, the end of the First World War. The reason for our choice of this interval is based only in part on the fact that the war curtailed mathematical production in the countries involved. It also marked a change in the nature of research in combinatorial group theory. New authors and new problems appeared after the war and the field acquired a life of its own. This will be the topic of Part II.

Of the eleven chapters of Part I, the first four after this introduction essentially follow the historical development, although Chapter I.5 deals with only one topic, the representation of groups by graphs. In these sections we have tried to reduce the technical difficulties to a minimum. We hope that they are easily accessible to anyone familiar with the basic facts of group theory. This may not be true for the lengthy Chapter I.6 which has been divided into six sections. It deals with the numerous group-theoretical investigations which appeared before 1918 but did not have an immediate influence on combinatorial group theory. They arose from many sources, e.g., the theories of arithmetically defined linear groups, of fuchsian groups, of Riemann surfaces, of differential equations, and even from the theory of finite groups, and became of interest in combinatorial group theory later (and, in many cases, much later). In Chapter I.6, we mention not only the original investigations but also some of their post-1918 sequels. This does not preclude our taking up the same topics again in a different context in Part II.

The prerequisites for Chapter I.6 are somewhat more rigorous than those for the other chapters. The opposite is true for Chapter I.7 where we give a survey of "the state of the art" at the end of the period under consideration.

Chapter I.8 does not deal with mathematics as such but with the ambiance of mathematical research during the period covered by Part I. The bibliographical notes which make up Chapter I.9 are mainly references to obituaries.

Chapter I.10 contains explanations of technical terms. Its purpose is twofold. The list of outdated terms should make access to the original pre-1918 literature somewhat easier for the modern reader, and the definitions of 22 current terms are meant to be an aid to the reader unfamiliar with the rudiments of combinatorial group theory. Also included in Chapter I.10 is a short list of standard notations.

Finally, Chapter I.11 briefly describes our sources. We have tried to consider all of the relevant papers published before 1918. We probably have not found all of them, although some gaps may be due to our interpretation of the word "relevant" and not to an oversight. We hope that the gaps are not too damaging.

Most of our quotations from the original research papers in foreign languages have been translated. We believe that in the case of mathematics, this procedure is defensible since here the danger of misrepresenting the original is extremely small.

Chapter I.2

The Foundation:
Dyck's Group-Theoretical Studies

The definition of a group G by a presentation, that is, by a system of generators and defining relations, is a particular aspect of the abstract definition of a group. As such, it has been discussed in Chapter 3, Section 4 of an extensive study by H. WUSSING [1969] on the genesis of the concept of an abstract group.

There can be no doubt that the paper by DYCK [1882] contains a decisive step in the definition of a group through a presentation. We shall not try to analyze the question of which concepts or ideas in DYCK's paper may have appeared earlier but shall confine ourselves to a statement of some of his results which are undoubtedly new and significant although the proofs may be open to criticism.

In the introduction to his paper, DYCK [1882] mentions the theory of automorphic functions and quotes papers by SCHWARZ, KLEIN, FUCHS, POINCARÉ, and SCHOTTKY in which groups appear as discontinuous (in today's sense) groups of geometric transformations where the elements of a group appear as the replicas of a fundamental region in a visible form. He then says that the geometric treatment of groups leads to a certain one-sidedness (*Einseitigkeit*) which had induced him to make a mistake in one of his earlier papers. He concludes the introduction with the following summary of the purpose of his paper:

> For the further development of the present group theoretical problems the analytic (combinatorial) formulation has to replace every geometric description. However, the primary geometric interpretation has produced certain viewpoints and it is the purpose of the present paper to develop both their geometric version and their analytic content.

The first section of DYCK's paper has the title *Definition of a Group G as the Starting Point of the Investigation*. We now quote this definition in translation:

> Let $A_1, A_2, A_3, \ldots, A_m$ be m operations of any kind which can be applied to an object J (identity) which, subsequently, will always be denoted by 1. Then these

A_i may always be considered as the *generating* operations of a group which will be obtained by applying all operations on our object J in iteration and combination.

The *most general* group with m generating operations will be obtained if we assume that the A_i do not have any periods and, in addition, are not connected mutually by any *relation*. We shall also consider the *opposite* operations of the A_i which we shall denote in the usual manner by A_i^{-1}. Then we obtain the infinitely many substitutions which belong to our group G if we apply first the operations $A_1, A_1^{-1}, A_2, A_2^{-1}, \ldots, A_m^{-1}$ to the identity, then apply to the thus resulting substitutions the same operations, and so on. Since we had assumed no reiation between the generating operations, the substitutions thus produced are all distinct from each other and *each of them can be obtained only by one completely determined process from the generating substitutions*. This is expressed by the formula

$$A_1^{\mu_1} A_2^{\mu_2} \ldots A_1^{\nu_1} A_2^{\nu_2} \ldots .$$

There are no comments on the ranges of the exponents ν, μ of the A_i in DYCK's paper, but later remarks clearly show that DYCK excludes the case where an A_i is preceded or followed by an A_i^{-1}.

Next, DYCK uses an artifice in order to avoid the need for using inverses by introducing an additional generator A_n and postulating that

$$A_1 A_2 A_3 \ldots A_m A_n = 1.$$

Next, he dedicates a section to what he calls "a geometric concretization (*Versinnlichung*) of the group G." Here he leans heavily on the theory of fuchsian groups. He constructs a $(m+1)$-gon P with circular arcs orthogonal to a fixed circle K and with vertices on K. The vertices are denoted by the symbols $a_1, a_2, \ldots, a_m, a_n$. Now we reflect P in all of its sides and continue to reflect its images in all of their sides and so on. This produces a tessellation of K with images of P. Assume that P is shaded and apply shading to all those images of P which arise from P through an even number of reflections. Consider now all the self-mappings of this tessellation of K which carry shaded images of P into shaded images. In the case under consideration, these self-mappings can be achieved through Moebius transformations mapping K into itself. Each mapping is uniquely determined by the image of P. These mappings then form a group isomorphic with the abstract group G constructed in the previous section of DYCK's paper. The proof is based on an argument which comes rather close to the use of the graph of the group (which, indeed, can be derived immediately from DYCK's tessellation). We quote:

The arrangement of the polygons in our net shows that, by starting from the polygon P and applying operations $A_1, A_2, \ldots, A_m, A_n$ exclusively in the *positive* sense, we can reach every other shaded polygon and *we can reach it through only one path* —apart from the insertion of paths $A_1 A_2 \ldots A_m A_n$, $A_2 A_3 \ldots A_n A_1$, etc. which are reducible to the identity.

The fourth section of DYCK's paper deals with the relationship between the group G with generators $A_1, A_2, \ldots, A_m$ and an arbitrary group $\overline{G}$ with generators $\overline{A}_1, \overline{A}_2, \ldots, \overline{A}_m$. He begins with the assumption that an operation

$$F(A_1, A_2, \ldots, A_m) = F$$

changes into the identity if the A_i are replaced by the $\overline{A}_i$ and observes that F and all its conjugates generate a subgroup H which commutes with all elements of G and is therefore, in a terminology ascribed to SOPHUS LIE, "distinguished." This group H consists of exactly those elements of G which become equal to 1 if the A_i are replaced by the $\overline{A}_i$. Here DYCK forgets the conjugates of F^{-1}. This oversight may be explained by his previously used construction which avoids inverses. DYCK then proceeds with the corresponding analysis of a group arising from G by assuming the validity of any number of relations between the generators, and he proceeds in the next section with the generalization which, in today's language, states that adjoining a relation for the generators of a group $\overline{G}$ produces a quotient group of $\overline{G}$. Finally, DYCK also gives a geometric interpretation of this construction of a quotient group of G, using a construction of its fundamental region in the tessellation corresponding to G with the help of coset representatives of a normal subgroup.

A streamlined account of DYCK's result may be found in BURNSIDE [1897a or 1911]. The paper by DYCK [1882] is mentioned even in the preface to this influential book. Altogether, it is one of the most widely quoted group-theoretical papers in the decades following its publication. Its importance is mentioned as late as 1935 in a historical article by G. A. MILLER and also in the survey by A. LOEWY [1910] on "algebraic group theory" which contains a particularly comprehensive and carefully written summary of what was then known in group theory outside of the theory of Lie groups.

DYCK's papers [1882 and 1883] contain much more than the foundation of the theory of group presentations. In particular, DYCK [1883] contains important contributions to the theory of permutation groups. But the most noteworthy effect of DYCK's paper of 1882 is probably that from then on the definition of a group through a presentation becomes a common feature in the literature. In many cases, a presentation is used as a very concise method of defining a group and a limited number of group properties; in particular, nonsolvability may be made obvious by a presentation. Probably for these reasons, presentations were derived soon after 1882 even for finite groups which can be defined easily by permutations or matrices. For instance, DYCK [1883] gives presentations for the simple groups of orders 60 and 168. BURNSIDE [1899] and FRICKE [1899] do the same for the simple group of order 504, and BURNSIDE [1897b] contains presentations for all of the symmetric groups. For later developments which show more essential uses of presentations of finite groups, see the monograph by COXETER and MOSER [1972]. Of course, the really significant role of presentations of groups appears in the theory of infinite groups. It will be described in later chapters of this book. For now, we shall merely try to put the two main results of the paper by DYCK [1882] into a historical perspective. Using current (1980) terminology, we may formulate these results as follows.

Proposition 1. *There exists a group G on m generators* $A_1, A_2, \ldots, A_m$ *such that every element of G can be written in exactly one manner in the form*

$$A_{i_1}^{e_1} A_{i_2}^{e_2} \ldots A_{i_l}^{e_l} \qquad or \qquad 1, \tag{1}$$

where

$$i_1, i_2, \ldots, i_l \in \{1, 2, \ldots, m\}, \qquad i_\nu \neq i_{\nu+1}, \qquad \nu = 1, \ldots, l-1$$

and where $e_1, e_2, \ldots, e_l$ *are nonzero integers.*

Proposition 2. *Every group* $\overline{G}$ *on m generators* $\overline{A}_1, \overline{A}_2, \ldots, \overline{A}_m$ *is a homomorphic image of G. We can obtain a group* $\overline{G}$ *by choosing an arbitrary set of expressions of the form* (1) *which we denote by* $F_1, F_2, \ldots, F_r$ *and postulate that*

$$F_\rho(\overline{A}_1, \overline{A}_2, \ldots, \overline{A}_m) = 1 \qquad (\rho = 1, 2, \ldots, r). \tag{2}$$

The kernel of the mapping $G \to \overline{G}$, $A_\mu \to \overline{A}_\mu$ ($\mu = 1, 2, \ldots, m$) *is then given by the products of the conjugates*

$$TF_\rho^{\pm 1}T^{-1}, \qquad T \in G,$$

where, in F_ρ, *we have replaced* $\overline{A}_\mu$ *by* A_μ.

One may say that DYCK's arguments for the validity of these propositions are convincing but not really rigorous. In particular, his geometric interpretation of Proposition 1 is merely intuitively clear but does not stand up under today's requirements for rigor. The proposition itself is accepted as self-evident, e.g., by BURNSIDE (see BURNSIDE [1911, p. 373]). It sounds plausible enough, but a purely algebraic proof is not entirely trivial. A particularly elegant one using permutations was given by SCHREIER [1927]. It is interesting to note that its use was avoided completely in the exposition of presentation theory given by DE SÉGUIER [1904] where, at the same time, interesting applications of this theory are introduced.

We know very little about DE SÉGUIER except that, according to WUSSING [1969] he was a private scholar without academic affiliation, something rather unusual for a very productive mathematician, particularly on the European Continent. His main work is a monograph on group theory which appeared in two volumes published separately in 1904 and 1912. They contain many original contributions of the author. The style of DE SÉGUIER is in sharp contrast to that of DYCK. There are no intuitive considerations (least of all, geometric ones) in DE SÉGUIER's work, and there is a tendency to be as abstract and as general as possible, although the general theorems are supplemented by many specific examples and investigations of special cases. DE SÉGUIER may have been the first algebraist to take note of CANTOR's discovery of uncountable cardinalities. His style is much more condensed than DYCK's or BURNSIDE's. This may have been the reason why his books appear to have been less influential than BURNSIDE's text in spite of the tremendous amount of material which they contain.

We shall be concerned here only with the first volume (published in 1904) of DE SÉGUIER's monograph. It begins with an introduction to set theory, quoting CANTOR [1895] and proceeds with the introduction of the concept of a semigroup with two-sided cancellation law. This entire concept is called "semigroup" by DE SÉGUIER, who coined the word (see p. 8 of his book). Next (pp. 15–16), DE SÉGUIER introduces generators $a_1, a_2, \ldots$ of a group with the remark that their number need not be countable. Now he sets himself the task:

> To find the general form for the relations which are consequences of a given system S of relations between the generators $a_1, a_2, \ldots$ (independent or not) if S defines a group. In this case a_i^{-1} has a meaning. ... Suppose S is given in the form $F_1 = 1, F_2 = 1, \ldots$ where the a_i^{-1} can appear in the relations.

He states:

> Then every one of the consequences of S can be put into the typical form
>
> $$\prod_{V,F} V^{-1} F^{\pm 1} V = 1$$
>
> through identical transformations (that is, by not assuming any relations except $a_i a_i^{-1} = a_i^{-1} a_i = 1$ between the formally distinct products).

Shortly thereafter (p. 17), DE SÉGUIER uses generators and defining relations to construct groups G which contain a normal subgroup A with quotient group $G/A = B$, where both A and B are given by generators and defining relations. The use of these concepts pervades much of his book.

We see that DE SÉGUIER, like DYCK, is in full possession of Proposition 2, although DE SÉGUIER's proof is more streamlined and more modern than DYCK's. This is partly due to the fact that the concept of a homomorphic mapping had, in the 22 years between the publications of DYCK and DE SÉGUIER, been widely used and fully understood. However, Proposition 1 does not appear at all in DE SÉGUIER's approach. Since he had introduced the concept of a semigroup, DE SÉGUIER did not have to assign a special role to free groups. His point of view (although not formulated in full detail) is essentially the same as that of M. DEHN who propagated it in his lectures and in discussions: A group arises from a set of symbols which appear in pairs a_i, a_i^{-1} by forming words with juxtaposition as an obviously associative composition. Equivalence classes of words with respect to certain rules which allow reversible changes of words are then the elements of a group.

Then the free groups appear merely as a special case where the rules of change guarantee the existence of an inverse for each group element. The fact that every mapping of a set of free generators a_i (without a_i^{-1}) onto a set of elements of any group G defines a homomorphic mapping of the free group into G is then completely obvious. What we called Proposition 1 is then what DEHN called the solution of the word problem for free groups. In DYCK's formulation, it appears as a theorem stating the existence of a free group. That is a point of view which has been maintained through much of the literature. Presentations of groups are then introduced as quotient groups of free groups. This procedure may have a slight technical advantage but there is indeed, as observed by DEHN in his lectures, no logical distinction between the concept of a free group and of that of a group given by any presentation. However, there is something here which has to be proved one way or another. And it is a remarkable fact that no algebraic proofs of Proposition 1 are given before 1926, and then they appear only as special cases of proofs concerning theorems for free products of groups. This development is very much in contrast to the corresponding development of the algebraic treatment of the concept of a group in general and particularly of the concept of a finite group. The book by WUSSING [1969] gives a careful account of the large number of papers dedicated to the axiomatic and algebraic conception of groups in general before 1919. A great many textbooks of algebra and, also, specifically, of group theory give a careful analysis of one-generator (i.e., cyclic) groups. But the first textbook of group theory which gives an explicit algebraic proof for the solution of the word problems in free groups is KUROSH [1944] (in Russian). In English, the first proof in a textbook appears in ZASSENHAUS, 1958, and even here it appears as a special case of the solution of the word problem for free products and is, incidentally, labeled as an existence theorem. The first combinatorial proof (in the sense of DEHN) appearing in a textbook seems to be the one in MAGNUS, KARRASS, and SOLITAR [1966]. The difference between the two conceptions of a free group expresses itself in the proofs. The existence proof starts with the normal form (1) of an element of the free group and proceeds with the demonstration that the combination of juxtaposition and subsequent free reduction of two such expressions satisfies the associative law. The combinatorial proof uses the semigroup of words and then shows that the normal form (i.e., the freely reduced word in an equivalence class of words) is independent of the specific reduction process used. Both approaches have been used by ARTIN in 1926 and in 1947a, respectively, in the more general framework of the existence of free products or, alternatively, the solution of the word problem for free products. Before 1926, the contents of Proposition 1 were accepted without hesitation, perhaps on the basis of the not entirely explicit geometric proofs which were available. The nature of these will emerge in the following chapters.

Chapter I.3

The Origin: The Theory of Discontinuous Groups

Even apart from DYCK's own testimony, there exists very strong evidence that the theory of discontinuous groups was the basis for the group theoretical studies of DYCK [1882]. DYCK was a student of F. KLEIN and, in 1882, was his assistant at the University of Leipzig. Shortly after the publication of DYCK's paper, there appeared two very important papers in the theory of discontinuous groups by the two leading mathematicians working in this field at that time. One of them was a long paper (61 pages) by H. POINCARÉ [1882] on fuchsian groups. In the introduction, POINCARÉ says that he had published sketches of his ideas and results earlier but that he would now try to give a systematic account of the theory. Almost simultaneously, there appeared a long (78 pages) paper by F. KLEIN [1883] which is particularly important for the theory of group presentations since it contains what is now known as KLEIN's theory of "composition of groups." This term was actually coined by FRICKE and KLEIN [1897, pp. 190–194]. KLEIN [1883] calls it *Ineinanderschiebung* ("meshing") of groups (p. 200). The composition of groups can be viewed as a solution of the word problem for the free product of groups which act in a certain manner on the points of a topological space. The most simple example, described by FRICKE and Klein, is the following one. Consider four disjoint circular disks D_1, D_1', D_2, D_2' in the complex plane. Let E be the common exterior of the disks and let A_1, A_2 be two Moebius transformations (i.e., bilinear self-mappings of the complex plane) such that A_1 maps the exterior of D_1 onto the interior of D_1' and A_2 maps the exterior of D_2 onto the interior of D_2'. Then for $\alpha_1 \in \mathbb{Z}$, $(\alpha_1 \neq 0)$, $A_1^{\alpha_1}$ maps E into the interior of either D_1 or D_1'. Since both D_1 and D_1' are outside of D_2 and D_2', a mapping

$$A_1^{\alpha_1} A_2^{\beta_1} \qquad (\alpha_1, \beta_1 \neq 0)$$

maps E into the interior of either D_2 or D_2'. A repetition of this argument shows that any mapping

$$A_1^{\alpha_1} A_2^{\beta_1} A_1^{\alpha_2} A_2^{\beta_2} \ldots A_1^{\alpha_m} A_2^{\beta_m}$$

with integers $\alpha_1, \ldots, \alpha_m, \beta_1, \ldots, \beta_m$ different from zero maps E onto a point set disjoint from E and, therefore, cannot be the identity. This proves the existence of a "most general" group on two generators in the sense of DYCK [1882] in a simple but nonalgebraic manner, and the argument can be generalized very easily to cover the free product of a countable number of cyclic groups. This was done, more or less explicitly, by FRICKE and KLEIN [1897]. These groups then also appear as discontinuous groups of Moebius transformations, acting on the complex plane. KLEIN's very simple idea has been refined and generalized in recent years for group-theoretical purposes; see, e.g., MASKIT [1965] and LYNDON and ULLMAN [1969].

The theory of discontinuous groups, even if we consider only groups of Moebius transformations, has exceedingly complex origins. They play an important role as tools used in the theory of algebraic functions of one complex variable (i.e., the theory of Riemann surfaces), uniformization, automorphic functions, the arithmetic theory of quadratic forms, the theory of algebraic extensions with a given Galois group (the theory of the equation of fifth degree), the theory of homogeneous linear differential equations (the hypergeometric equation) enter into it, and the geometric ideas which led to the development of non-euclidean geometry, including differential geometry. However, we have to confine ourselves here to the relationship between the theory of discontinuous groups and the theory of presentations of groups. On the one hand, presentations of discontinuous groups appear in large numbers in the literature before 1914, particularly in FRICKE and KLEIN [1897]. (A survey of these results may be found in MAGNUS [1974a]). On the other hand, these presentations are not really used very much, and they may be considered more as a convenient shorthand for the definition of the groups than as an aid for the investigation of their properties. In particular, nowhere does there appear a systematic attempt at deriving the presentation of a subgroup—even of a normal subgroup of finite index—from the presentation of the whole group. And this is true in spite of the fact that there is much interest in the subgroups of at least some discontinuous groups, in particular, of the elliptic modular group $\mathrm{PSL}(2, \mathbb{Z})$. (A brief description of some of the work dealing with the study of subgroups will be given in Chapter I.6.)

The reasons why the early theory of discontinuous groups did not stimulate the development of the theory of group presentations are quite obvious: The discontinuous groups are not primarily given by presentations. They are defined either by a set of generating elements which describe circle-preserving conformal self-mappings of the complex plane or as a set of 2×2 matrices, subject to arithmetic conditions (e.g., having as entries algebraic integers satisfying certain relations and lying in a given field). We shall call these two situations, respectively, the *geometric* and the *arithmetic* case. The simplest example for the geometric case is the one given above when explaining KLEIN's composition theorem. The simplest example for the arithmetic case is again PSL$(2, \mathbb{Z})$.

In both cases, the presentation of the group is derivable from the construction of the fundamental region, and the derivation is easy if the fundamental region is two dimensional, i.e., in the complex plane. In the geometric case, the main problem is to determine generating mappings with a prescribed fundamental region and prescribed correspondences between the circular arcs of its boundary under the action of the generators. This is a more or less difficult geometric problem which involves no group-theoretical methods. In the arithmetic case, one has to find the generators first. This is usually nontrivial and can be very difficult as in the cases treated in FRICKE and KLEIN [1897, Part 3] or in the paper by BIANCHI [1892] which will be described in more detail in Chapter I.6. The methods used in these papers can be arithmetical or geometric or both. The same is true for the construction of the fundamental region. Once this has been completed, the defining relations for the group are rather easy to find in the case of a two-dimensional fundamental region. (In three or more dimensions, this problem becomes much more difficult.) There are a few papers before 1914 which became important much later for some aspects of combinatorial group theory. They will be discussed briefly in Chapter I.6. However, the problems occurring in these papers could not and, of course, did not stimulate research in this direction. There is only one problem in the theory of discontinuous groups which is widely discussed in the pre-1914 literature where a development of methods of combinatorial group theory would have been both useful and naturally suggested by the nature of the problem. It is the task of finding all subgroups or, at least, all normal subgroups of finite index in a group given by a finite presentation, and, sometimes, the additional task of obtaining information about the possible finite quotient groups. The first mathematician to deal with this problem in a general and abstract way was K. REIDEMEISTER in 1926. His paper is entitled *Knots and Groups*. It is no coincidence that this is a topic from topology. The motivation for the development of combinatorial group theory beyond its rudiments had come from topology long before REIDE-MEISTER. We shall give the details in the next chapter.

Chapter I.4

Motivation: The Fundamental Groups
of Topological Spaces

In a long paper entitled *Analysis Situs*, Poincaré [1895] introduces the concept of the fundamental group of a topological space. He begins with a heuristic introduction, using functions F_i ($i = 1, \ldots, \lambda$) (not necessarily single valued) on a manifold defined by equations between coordinates x_k ($k = 1, \ldots, n$) and he assumes that these functions satisfy certain differential equations,

$$dF_i = X_{i,1} dx_1 + \cdots + X_{i,n} dx_n,$$

where $X_{i,k}$ are known single-valued differentiable functions of x_k and F_i which satisfy certain integrability conditions. Then he considers the transformations of F_i which result if one traces their values along a closed loop. These transformations form a group g and Poincaré demonstrates, or at least recognizes, that all possible such groups g are homomorphic images of a single group G, the fundamental group.

Poincaré looked upon group theory as an aid to topology and he was successful in this respect by showing that the fundamental group indeed determines the topological invariants of a space which had been found by Betti. He also saw that the definition of the fundamental group as a characteristic of a topological space leads to a more sophisticated classification of spaces than had been available before. His most noteworthy contribution in this respect is the construction of a three-dimensional space for which the Betti number and the torsion coefficients (which had been introduced by Poincaré [1904]) are the same as for the closed three-dimensional spherical space but which has as fundamental group Γ a perfect group which, in turn, has the group of the icosahedron (i.e., A_5) as a quotient group. This result is contained in Poincaré [1904, pp. 493–496, Vol. 6 of his collected papers]. At the very end, Poincaré raises the question which is now known as the "Poincaré Conjecture":

> *Est-il possible que le groupe fondamental de V se réduise à la substitution identique, et que pourtant V ne soit pas simplement connexe?* ("Is it possible that the

fundamental group of V consists only of the identity, and that yet V is not simply connected?")

This problem is still (1980) open. But at least it can now be transformed into a fully algebraic (essentially group-theoretical) version; see BIRMAN [1973], where other algebraic versions are also quoted. See also LYNDON and SCHUPP [1977, p. 195].

POINCARÉ's papers on topology are difficult to read for a variety of reasons. One of them is purely technical: He writes both abelian and nonabelian groups with addition as composition, passing from one case to the other without much warning. But the main difficulty arises from the fact that POINCARÉ uses construction methods for topological spaces which are taken from intuitive generalizations of ideas and results appearing in the theory of fuchsian groups, of Riemann surfaces, and of differential equations without ever trying to separate intuition from proof or to clarify his assumptions. It is clear from the construction of the space with a perfect fundamental group that this group is given by a finite presentation. But apart from this example, one cannot say that POINCARÉ used group-theoretical methods in any decisive manner. This remark is illustrated by the fact that he did not observe that his torsion invariants (in addition to the BETTI number) of a space are indeed computable by abelianizing the fundamental group. This result was found by TIETZE [1908] in a paper of 118 pages which is of importance not (like POINCARÉ's work) because of the introduction of new ideas but because of the careful analysis of the underlying assumptions and methods, because of a serious effort to separate intuitive arguments from genuine proofs, and finally because of its group-theoretical contributions.

TIETZE begins with a definition of manifolds (of all finite dimensions) through cell complexes which he also calls *Schemata* (schemes). He mentions various sources for this construction which defines a manifold (actually a space) in terms of finitely many data. The sources include papers by DYCK, by POINCARÉ (whose work he acknowledges as the basis of his entire paper), and unpublished lectures by WIRTINGER. But TIETZE [1908] also states that the systematic development of the construction he uses agrees with the one given in DEHN and HEEGAARD [1907] which had appeared a year earlier.

In Section 12 of his paper, TIETZE [1908] introduces the fundamental group of a manifold on the basis of his construction and shows that the fundamental group has a finite presentation. In Section 13, TIETZE shows that the fundamental group is a topological invariant. That is, he proves that two homeomorphic *Schemata* define isomorphic fundamental groups. For this purpose, he needs a purely group-theoretical result which he had proved in Section 11 and which is associated with his name even today. It is the theorem that any two finite presentations of a group can be carried into each other by applying finitely many elementary invertible transformations (also called "Tietze transformations") a finite number of times.

Section 11 of TIETZE's paper is prefaced by a brief summary of the concepts of generators and defining relations. TIETZE then remarks:

> One observes at once that it may happen that two groups are isomorphic although they are defined by using different systems of generators and defining relations. ... However, neither the general problem of characterizing the totality of all abstract ways of generating a given group nor even the special problem of finding a method to decide whether two groups given by their presentations are isomorphic have been solved.

TIETZE then introduces exponent sums for the generators in the defining relations of a group, considers the matrix of these exponent sums and its elementary divisors, points out that these are essentially the invariants of a finitely generated abelian group arising from the given group by abelianizing it, and finally proves that, for isomorphic groups, the invariants of the abelianized groups coincide. This leads him to the discovery, mentioned before, that POINCARÉ's torsion coefficients can be computed from the presentation of the fundamental group. He includes in this discourse examples of pairs of nonisomorphic groups with identical invariants of the abelianized groups.

At the end of Section 14 (p. 80) in Tietze [1908], he summarizes his results about the role of the fundamental group as follows.

> ... therefore the fundamental group of an orientable (*zweiseitige*) closed three-dimensional manifold contributes more to its characterization than all the presently known topological invariants taken together. However, this statement has to be restricted to some extent. Whereas it is always possible to decide the coincidence of two sequences of numbers, the question whether two groups are isomorphic is not always answerable. In contradistinction to the other topological invariants, the coincidence or noncoincidence of the fundamental groups of two manifolds is not always decidable.

The last section of TIETZE's paper contains the construction of two nonhomeomorphic manifolds both of which have a fundamental group of order 5.

The group-theoretical parts of Tietze [1908] are remarkably lucid. The proofs are very clear and obviously correct. The appraisal of the topological parts would go far beyond the scope of this book. It should, however, be noted that TIETZE gives an abundance of references and acknowledges his dependence on POINCARÉ's work at every step. In fact, some parts of his long paper may be read as a clarifying commentary on those by POINCARÉ.

TIETZE's result on the equivalence of finite presentation and his test for isomorphism of groups based on the isomorphism of the abelianized groups are the first theorems beyond those stated by DYCK and DE SÉGUIER in combinatorial group theory. Both theorems are proved group theoretically and both arise from POINCARÉ's discovery that group theory is applicable to topology. Also, TIETZE's results are motivated by an observation of a specific difficulty arising when working with groups which are merely given by a presentation.

The four papers by DEHN [1910, 1911, 1912, 1914] deepen and continue the work of TIETZE in a remarkable manner. DEHN, too, acknowledges POINCARÉ's discovery of the fundamental group as the motivation for his work. He recognized that the difficulties of combinatorial group theory start at a much lower level than that of the isomorphism problem as stated by TIETZE.

He contributes solutions to the group-theoretical problems raised by him in some important cases. And he solves a topological problem raised by TIETZE [1908, p. 98] by investigating the fundamental group of a particular space. But, unlike TIETZE, he does not use algebraic methods. His algebraic results are proved by using arguments from the theory of one-dimensional complexes. We proceed with an account of the details.

DEHN [1910] is a paper with the title *On the topology of the three-dimensional space*. However, it starts with a chapter on groups defined by a finite presentation. The following paper, DEHN [1911] is dedicated entirely to groups given by a presentation. We shall discuss these papers together, referring to them, respectively, as the "first" and the "second" paper.

The second paper begins with a formulation of three fundamental problems:

1. The Word Problem (called *Identitaetsproblem* by DEHN). Let an arbitrary element of the group be given through its buildup in terms of the generators. Find a method to decide in a finite number of steps whether this element equals the identity element or not.

2. The Conjugacy Problem (called *Transformationsproblem* by DEHN). Any two elements S and T of the group are given. Find a method to decide whether S and T are conjugate, i.e., whether there exists an element U of the group which satisfies the relation

$$S = UTU^{-1}.$$

3. The Isomorphism Problem "Two groups are given. To decide whether they are isomorphic or not (and also whether a given correspondence is an isomorphism)."

The first two problems are stated in both papers, but emphasized in the second. The third problem appears only in the second paper but, as mentioned before, had already been formulated (without emphasis) by TIETZE [1908]. DEHN [1911] gives a topological motivation for the formulation of these three problems. We quote:

Every knotted curve in space requires for its complete topological characterisation the solution of the three problems in a special form. To every curve K there corresponds an infinite group G_K defined in the manner described above (i.e., through a finite presentation). K is not knotted if and only if G_K is abelian. This implies the solution of a word problem. To every other curve in space there corresponds, with respect to G_K, a definite element of G_K. Two curves in space can be deformed continuously into each other without penetrating K if and only if the corresponding elements of G_K are conjugate. Finally, the question whether a given curve K' can be deformed continuously into K without penetrating itself requires the solution of the third problem for $G_{K'}$ and G_K.

This passage requires an inordinate amount of comments.

In his first paper, DEHN had proved a purely topological result which is still known as "Dehn's Lemma" and which implies that the group of a knot is abelian and therefore cyclic if and only if the knot is isotopic in three space with a circle. There was a mistake in the proof of the lemma which was pointed out in a letter dated April 22, 1929 from H. KNESER to DEHN. The lemma, however, is true and was proved by PAPAKYRIAKOPOULOS in 1957, five years after DEHN's death. But all of this does not really affect the importance of DEHN's paper for the development of combinatorial group theory. It is mentioned here merely because the asphericity of knots appears in DEHN's introduction to his second paper. Even without DEHN's lemma, it is clear that a knot cannot be isotopic with a circle in three space if the group of the knot is nonabelian.

In part, the importance of DEHN's work for combinatorial group theory is certainly based on his discovery that a presentation of a knot group can be read off from a projection of the knot onto the euclidean plane provided that this projection is sufficiently regular. DEHN associates a generator with each one of the finitely many regions into which the knot projection decomposes the plane and associates a defining relation with each one of the finitely many double points of the projection which have to be characterized as over- or under-crossings. His defining relations involve three or four generators or their inverses. DEHN's method also works if one wishes to study linkages, that is, the embedding of more than one closed curve into three space. We thus have a group defined by a finite easily computable presentation associated with each knot in a characteristic manner, and since LISTING [1848], knots had constituted a much investigated class of topological objects. This discovery obviously motivated DEHN. But it is worth noting that he was not the only mathematician, and not even the first, who made this discovery. WIRTINGER [1905] in a lecture delivered at a meeting of the German Mathematical Society (Deutsche Mathematiker-Vereinigung), explained that the algebraic singularities of an analytic function of two complex variables define topological structures, namely, a system of knotted or linked curves in three space which is obtained by studying the intersection of the corresponding algebraic surface (a manifold of two real dimensions embedded in a space of four real dimensions) with a hypersphere which is sufficiently small and whose center is the singular point of the surface. And WIRTINGER also had given a method of defining the fundamental group of a knot and had shown how to read it off from a projection of the knot into the Euclidean plane. WIRTINGER's method is different from DEHN's because it associates the oriented segments of the projection of the knot with the generators of the group, but it is equally simple and has, in fact, been preferred by later authors in monographs on knot theory, e.g., REIDEMEISTER [1932a] who gives the same reference to WIRTINGER [1905] as we do. Now it is true that WIRTINGER never published his results and ideas. We know about them from the paper of one of his students, BRAUNER [1928] which appeared much later. But although the reference to WIRTINGER's lecture consists only of one line, its contents must have been known widely. In particular, TIETZE [1908, pp. 96, 105] considers the same knot (trefoil or clover knot) as DEHN and gives a presentation of its group and, a few lines later, quotes WIRTINGER [1905]. There is no indication known to us that DEHN was aware of WIRTINGER's ideas, but in view of the great importance of algebraic geometry at that time, one might have guessed that WIRTINGER's discovery would have provided a particularly strong motivation for the investigation of knot groups. However, it was DEHN who developed group theory for the purposes of topology. There are two outstanding applications.

In his first paper, DEHN constructs infinitely many "Poincaré spaces," that is, closed three manifolds with a perfect nontrivial fundamental group. He is able to write down presentations for such groups explicitly by constructing his spaces through the amalgamation of the surfaces of a knotted and an unknotted torus. The fundamental group is then a quotient group of the knot group associated with the knotted torus. To arrange it so that the abelianized fundamental group of the space is trivial is a very elementary problem. The difficulty which arises is to prove that the group itself is nontrivial. This is indeed part of a word problem. DEHN not only shows that the fundamental groups which he obtains are nontrivial, but he also proves that they are, with one exception (the one discovered by POINCARÉ [1904], of infinite order. For this purpose, he uses the graph-theoretical method developed by him which will be discussed in the next chapter.

The second important problem which DEHN [1914] solved with group-theoretical methods is one which had been already posed by TIETZE [1908, p. 97]. DEHN proves that the trefoil (or clover) knot cannot be deformed continuously without self-penetration into its mirror image. His proof is based on the construction of all automorphisms of the group of the knot. (That topological self-mappings of the space induce automorphisms of the fundamental group had already been recognized by TIETZE [1908, p. 90].) This is probably the most famous of all of DEHN's papers, in spite of the fact that already his *Habilitationsschrift* (DEHN [1901]) contained the solution of the third of the 23 famous problems posed by HILBERT at the International Congress of Mathematicians in Paris in 1900.

DEHN [1911] observed that a finitely presented group may have subgroups which are not finitely presented. As an example, he takes the subgroup of a free group on two free generators S_1, S_2 which is generated by the conjugates of S_1.

DEHN [1911] also solved both the isomorphism problem and the conjugacy problem for finitely presented groups which have the property that each generator appears at most twice in the defining relations. The term *free generator* (meaning a generator not appearing in any defining relation) appears here for the first time. The proofs are based on arguments taken from the theory of discontinuous groups of non-euclidean planar motions and from topological considerations. The groups eventually appear as fundamental groups of two-dimensional manifolds with singularities.

DEHN [1912] once more took up the word problem and the conjugacy problem for the fundamental groups Φ_g of orientable closed two-dimensional manifolds. These are given by generators a_i, b_i, $i = 1, \ldots, g$ and a single defining relation $R = 1$ where

$$R = a_1 b_1 a_1^{-1} b_1^{-1} a_2 b_2 a_2^{-1} b_2^{-1} \cdots a_g b_g a_g^{-1} b_g^{-1}.$$

Now DEHN [1912] proved the following theorem.

Assume that $g > 1$, and let W be a freely reduced word in the generators a_i, b_i. If $W = 1$, then there exists a subword W_0 of W of length at least $2g + 1$ which is also a subword of a cyclic permutation of $R^{\pm 1}$.

This theorem obviously solves the word problem for the groups Φ_g for $g > 1$. (For $g = 1$, the theorem is not true, but then a near-trivial solution of the word problem exists.) The point here, however, is not the solution of the word problem for these particular groups. DEHN had done this already, in a cumbersome fashion, in his 1911 paper, using non-euclidean geometry and the fact that the groups Φ_g can be considered as discontinuous groups of planar non-euclidean motions with a fundamental region which is a regular $4g$-gon in the non-euclidean plane. What is essential here is that Dehn had found an algebraic solution of the word problem (although not an algebraic proof). This algebraic solution has the following simple characteristics.

Take a word W in the generators. Reduce it freely. If $W = 1$ in the group, then a subword W_0 of W also appears in a cyclic permutation of the freely reduced relator or its inverse and the number of symbols (= generators and their inverses) appearing in W_0 is greater than one-half the number of symbols in the freely reduced relator. Replace W_0 by the inverse of the remaining part of the relator. This leads to a shorter word W'. Reduce W' freely and continue in the same manner. Then and only then will we have $W = 1$ if our procedure terminates with the empty word.

This procedure is now called *Dehn's algorithm* for the solution of the word problem of the groups Φ_g. The term was coined by MAGNUS and appears for the first time in the title of the Ph.D. thesis of GREENDLINGER [1960a]. There it is shown that Dehn's algorithm is applicable to a large class of groups with finite presentation, and GREENDLINGER's proofs as well as his results are purely algebraic. They contain DEHN's result as a very special case.

DEHN [1912] actually has sharper results than the one quoted (which, in content although not in formulation, is identical with his theorem on p. 415). Also, DEHN [1912] gave a similarly simple purely algebraic solution for the conjugacy problem of the groups Φ_g. In his previous paper (DEHN [1911]) he had done this using non-euclidean geometry and phrasing the solution in geometric terms. Again, these results of DEHN have been strongly generalized with purely algebraic proofs, by GREENDLINGER [1960b] and there exists an extensive literature on Dehn's algorithm subsequent to GREENDLINGER's paper which we cannot discuss here. However, we have to say something about DEHN's proofs. They are nonalgebraic. The method used by DEHN is essentially that of his *Gruppenbild*, which will be discussed in detail in the next chapter.

Chapter I.5

The Graphical Representation
of Groups

The title of this chapter is part of the heading of Chapter 19 of BURNSIDE's *Theory of Groups of Finite Order*, 2nd edition, which appeared in 1911. It comprises the description of a group by a tessellation of the sphere, the euclidean or the non-euclidean plane by replicas of a fundamental region of the group which, in this case, must act discontinuously on the manifold in question and also the one-dimensional complex which can be associated with any group presentation and is known under the names "Cayley diagram," "colour group," *Gruppenbild*, and "graph of a group." We shall use the last term here.

It is obvious that a tessellation of a two manifold with replicas of a fundamental region of a group acting on the manifold leads immediately to the graph of the group by taking the dual complex of the one provided by the tessellation, and this is the way in which BURNSIDE [1911] introduces it, although he follows it up with an abstract definition which shows that the graph of a group is independent of the existence of a discontinuous action of the group on a two manifold. Before discussing the impact of the underlying idea, we shall first give a historical account of the concept of the graph of a group and of its applications, leaving out most of those papers which deal exclusively with finite groups. For these, BURNSIDE [1911] or COXETER and MOSER [1965] is a good reference.

CAYLEY [1878b] begins his paper with the words

> I recapitulate the general theory so far as is necessary in order to render intelligible the quasi-geometrical of it which will be given.

Here "it" refers to the title of the paper, which is *On the theory of groups*. CAYLEY briefly summarizes the definition of a group, sets up the group table for the group A of even permutations of four symbols, and observes that it is not necessary to introduce 12 distinct letters for the 12 elements of the group but that they can be written in terms of two elements α, β as

$$1, \alpha, \alpha^2, \beta^2\alpha\beta^2, \beta\alpha^2, \beta^2\alpha, \beta^2, \beta\alpha^2\beta\alpha^2, \beta\alpha, \beta, \beta^2\alpha^2, \beta^2\alpha\beta^2\alpha.$$

Next, CAYLEY considers what he calls a particular "hemihedron," namely, a cube whose "summits" have been truncated to obtain a polyhedron of eight triangular faces and six square faces which he then represents by the diagram below, where, obviously, the polyhedron is obtained by identifying the pairs of edges whose endpoints appear twice with the same labels a, j, k. CAYLEY continues:

> ... a diagram, the lines of which were red and black, and they will be thus spoken of, but the black lines are in the woodcut continuous lines, and the red lines broken lines: each face indicates a cyclical substitution, as shown by the arrows. The figures should be in the first instance drawn with the arrows, but without the letters; but I have in fact affixed them in such wise that the group given by the diagram, as presently appearing, may (instead of being any other equivalent group) be that group which contains the before-mentioned substitutions
>
> $$\alpha = abc \cdot def \cdot ghi \cdot jkl$$
>
> and
>
> $$\beta = ajg \cdot bif \cdot cek \cdot dhl.$$

> Observe that in the diagram, considering the lines to be drawn as shown by the arrows, there is *from* any given point whatever only one black line, and only one red line...
>
> The diagram has the property that every route, leading from any one letter to itself, leads also from every other letter to itself.
>
> ...*it is in virtue of this property* that the diagram gives a group.

CAYLEY then discusses his special example further and observes that his construction can be generalized and that diagrams of his type with certain

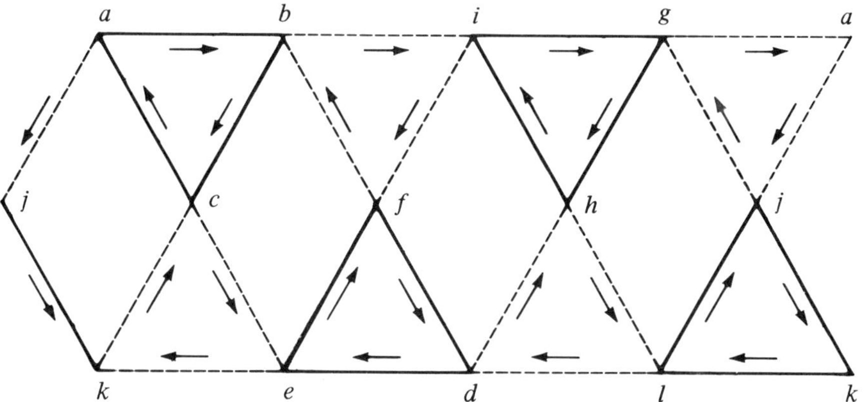

Figure 1. Cayley's diagram for the alternating group A_4 of order 12.

properties give rise to a group. The converse is not difficult to establish, and we now quote the definition given by BURNSIDE [1911, pp. 424–425]:

> Suppose now that...we have a diagram of N points connected by $\frac{1}{2}N(N-1)$ coloured directed lines satisfying the following conditions:
>
> (i)   all the lines of any colour have either (a) a single arrowhead denoting their directions or (b) no arrowhead, in which case the line may be regarded as equivalent to two coincident lines in opposite directions;
> (ii) there is a single line of any given colour leading to every point in the diagram, and a single line of the colour leading from every point: if the colour is one without arrowheads, the two lines coincide;
> (iii) every route which, starting from some one given point in the diagram, is closed, i.e., leads back again to the given point, is closed whatever the starting point.
>
> Then, under these conditions, the diagram represents in graphical form a definite group of order N.

Later on, BURNSIDE [1911, p. 426] observes that we may leave out all lines with a fixed color if this process does not make the diagram disconnected, and he concludes:

> The simplest diagram that will represent the group will be that which contains the smallest number of colours and at the same time connects all the points... It may be noticed that this simplified diagram can be actually constructed from the previously obtained representation of the group by the regular division of a surface...

Instead of BURNSIDE's definition of a colour group, we could have used the one given by MASCHKE [1896] which is practically the same. Both appear to be no more than streamlined versions of CAYLEY's idea. We wish to emphasize that DEHN's definition of the *Gruppenbild* is based on a different approach. CAYLEY, MASCHKE, and BURNSIDE assume the group to be known and construct the graph which then allows them to recognize generators and defining relations. DEHN [1910] starts with a presentation of a group G, given in terms of finitely many generators a_ν, $\nu = 1,\ldots,k$ and finitely many defining relations $S_1 = 1,\ldots,S_m = 1$ and then describes a procedure for the construction of the *Gruppenbild*. We quote the first paragraph of this construction:

> We construct the following one-dimensional complex (*Streckenkomplex*) C^G, associated with G: Let
>
> $$S_1 \equiv a_{k_1}^{\epsilon_1} a_{k_2}^{\epsilon_2} \ldots a_{k_l}^{\epsilon_l} \qquad (\epsilon_i = +1 \text{ or } -1).$$
>
> We choose a point Z and a circle (closed curve) K^1 with l vertices $Z, P_1,\ldots,P_{l-1}$ and we denote ZP_1, by $+a_{k_1}$ or by $-a_{k_1}$, $P_1 P_2$ by $+a_{k_2}$ or by $-a_{k_2}$ and so on according to the values $+1$ or -1 of $\epsilon_1, \epsilon_2,\ldots$. Next, we take a second circle K^2 the edges of which, starting from Z and passed through in a given direction, will be noted by $+a_{k_2}$ or $-a_{k_2},\ldots, +a_{k_l}$ or $-a_{k_l}, a_{k_1}$ or $-a_{k_1}$. (The decision about the signs is again made by considering the values of the ϵ_i.) In this manner we

continue until we have exhausted all cyclic permutations of S_1. We then will have l circles attached to Z the edges of which, in the given orientation, have been given the notation described above. If we reverse the orientation the edges will always be affixed with the opposite sign. Suppose now that we have two edges ZP and ZQ which, with Z as the initial point, carry the same notation. We then shall identify P and Q as well as ZP and ZQ, and we shall continue with this process until the notations of all edges starting at Z are distinct. We shall carry out the same process for all other points of all of our circles and we shall continue until from every point there will emerge only edges with distinct notations. We shall denote the complex thus obtained by $\bar{C}^1$, and we shall call Z its center.

The construction goes on, taking into account the remaining (if any) defining relations for G. Although, in general (i.e., for infinite groups), the construction will require infinitely many steps, DEHN notes that the following must be true: Consider any two distinct points P and Q, obtained after r_0 steps of construction. If P and Q are still distinct after $r > r_0$ steps of construction, for a sufficiently large r, they will stay distinct after $s > r$ steps of construction, no matter how large s is. DEHN also observes that he has merely proved the existence of a *Gruppenbild* but has not given a method of constructing it in a finite number of steps. He also observes that the construction of the *Gruppenbild* implies the solution of the word problem for G. (Obviously, the two problems are identical, but DEHN does not mention this.)

Finally, DEHN [1910] reproduces a drawing of the graph of the finite group with generators a_1, a_2 and defining relations

$$a_1^5 = 1, \qquad a_2^3 = 1, \qquad a_1 a_2 a_1 a_2 = 1$$

and remarks in a footnote:

> The assignment of the group of the icosahedron to this complex is known. (See Maschke, *Am. Journ.* 1896.) Altogether, the *Gruppenbild* is nothing really new in the case of finite groups. It is closely related to Cayley's "Colour Diagram".

It is hard to understand how DEHN could not have observed that his *Gruppenbild* is identical with CAYLEY's colour diagram, and not only in the case of finite groups. It is, however, true that CAYLEY's idea had, before DEHN, been applied mainly to finite groups where it leads to the definition of the genus of a group as the smallest genus of a closed orientable two-dimensional manifold on which a graph for the group can be drawn without any edges intersecting at points other than the vertices.

Before discussing the use DEHN made of the graph of a group, we shall briefly mention its effect on the theory of finite groups. It certainly has been thought of as something very attractive and pleasant. The book by BURNSIDE [1911] has a "colour group" (that of the octahedral group or Σ_4) as a frontispiece. G. A. MILLER, who was, in practically all of the 359 papers reprinted in his collected works, concerned only with finite groups, writes as late as 1935:

> These diagrams are useful for expository purposes but do not seem, up to the present time, to have led to any new theorems in group theory.

In a way (if one excludes infinite groups, which MILLER did not care about and did not know very much about) this is true even today (1980). The promise of obtaining a new and interesting classification of *finite* groups according to their genus has not been fulfilled. One may mention here that the fact that the simple group G_{168} of order 168 has genus 3 is important for the theory of equations of degree 7 with Galois group G_{168}. It is impossible to develop the theory of these equations in the same manner as the theory of the general equation of fifth degree with Galois group A_5 since the automorphic functions belonging to a Riemann surface of positive genus are not simply the rational function of a single automorphic function. See FRICKE [1926, pp. 211–240]. However, this result may be too special to carry much weight here.

Coming back now to DEHN, his use of the graphs of *infinite* groups has indeed contributed heavily to their theory. We start with DEHN [1910], where we find the construction of the graph of the group of the trefoil knot which is given by four generators C_1, C_2, C_3, C_4 and the relations

$$C_1 C_4^{-1} C_2 = C_2 C_4^{-1} C_3 = C_3 C_4^{-1} C_1 = 1.$$

DEHN does not use his recipe for the construction but presents the finished result which is surprisingly simple and, at least intuitively, completely convincing. The fact that the group is not abelian becomes obvious, and all the information needed about the group in both DEHN's papers of 1910 and 1914 can be derived from the graph. It should be noted that the group of the trefoil knot has as its center an infinite cyclic group whose quotient group is the well-known modular group PSL(2, $\mathbb{Z}$) defined by two generators a, b and relations $a^2 = b^3 = 1$. DEHN could have used this fact for the construction of the graph since the graph of the modular group can be read off from the well-known tessellation of the non-euclidean plane by its fundamental region. However, DEHN's construction is a direct one and very simple. In a footnote, DEHN [1914] also mentions that a Ph.D. student of his had constructed "the interesting *Gruppenbild*" of the group of the figure eight (or Listing's) knot. All we know about this (through oral tradition) is the name of the student, FRITZ KLEIN. It is likely that he, like DEHN's student GIESEKING, perished in World War I, which took a terrible toll of the youth (and, particularly, the academic youth) of Europe.

DEHN [1911] had prefaced his chapter on the fundamental groups of closed surfaces with the remarks:

> It would not be proper to say that, in the case of fundamental groups of closed surfaces, pure group theory can be of much help to topology. On the contrary, the solution of topological problems has a stimulating effect on group theory.

In this paper, DEHN deals with the word and conjugacy problems of the fundamental groups of closed two-dimensional manifolds on the basis of a tessellation of the non-euclidean plane by the fundamental regions for discontinuous groups of Moebius transformations which are faithful representations of the fundamental groups. Although the existence of such representations was known for quite some time (and had been recorded, for instance, in the monumental work of FRICKE and KLEIN [1897]), no explicit formulas for such a representation were known, and these fill the first chapter of the Ph.D. thesis of DEHN's student GIESEKING which appeared in 1912. DEHN [1912] refers to GIESEKING's paper. But in the meantime, DEHN had observed that he did not need non-euclidean geometry at all to solve the word and conjugacy problems for these groups. All he needed were topological properties of the graph of the group (which is simply the dual of the tessellation of the non-euclidean plane by non-euclidean polygons). The properties in question are as follows: The graph consists of $4g$-gons ($g > 1$) and at each vertex exactly $4g$ of these meet. DEHN [1912] observes that these assumptions can be weakened: It is enough to assume that the graph consists of polygons with at least seven vertices and that at each vertex at least four of the polygons meet. The fact that any two polygons must not have more than one edge in common is tacitly implied. That the number seven is, in a sense, optimal emerges from the paper by GREENDLINGER [1960b].

Independent of the specific results derived by DEHN with the help of his *Gruppenbild*, there remains the question whether the graphical representation of groups as initiated by CAYLEY provides more than an aid which, in the words of MILLER [1935] is "useful for expository purposes." After all, DEHN's results can be derived algebraically and, in cases more complicated than those studied by DEHN, the algebraic methods are simpler because the description of the graph of a specific group can become very frustrating. Already in the case of the group of Listing's knot (the "figure-eight knot"), nobody has ever tried to reconstruct the graph of the group which DEHN's student FRITZ KLEIN had found. The fact that the group is an infinite cyclic extension of a free group of rank 2 can be established algebraically with great ease and allows a straightforward computation of its group of automorphism (MAGNUS [1931]). And the times of an intuitive use of topology are past. DEHN used to say that he had always known the result in the paper of SCHREIER [1927a], that subgroups of free groups are free because, "after all, subgraphs of a tree are trees." But the orderly derivation of the fact that the topological statement is true and that it implies the group-theoretical one requires some work, and, on the other hand, SCHREIER's algebraic proof is not very difficult.

We believe that in spite of all objections of this type, DEHN was right. And he was right not only in the general sense of his remark that topological problems and, we may add, concepts have a stimulating effect on group theory. DEHN's insistence on the importance of the *Gruppenbild*, which was particularly pronounced in oral discussions, had as its basis the insight that the graphical representation of a group was the adequate expression of an aspect of its very nature. Graphs have been used in combinatorial group theory, at least intermittently, ever since DEHN [1910] had introduced them as a tool, and the very paper by SCHREIER [1927a] mentioned in DEHN's quotation above contains an interesting extension of DEHN's construction, the coset graph of a subgroup. But the most striking confirmation of DEHN's evaluation is probably the paper by STALLINGS [1968]. The basic theorem proved there states: "If a finitely presented torsion-free group has infinitely many ends, the group is the free product of two nontrivial factors." Now the number of ends of an infinite group can be defined as follows (HOPF [1944]): Remove a finite number of edges from the graph of the group. Count the number of disjoint infinite connected parts of the remainder of the graph. The least upper bound for this number is the number N of ends of the group. N can have only one of three values, namely, 1, 2, or ∞.

The point to be made here is that this is such an extremely simple definition when phrased in topological terms. Any possible translation into algebraic terms appears to be clumsy and opaque. Of course, one may say that all that matters about a definition is its precision. But this was certainly not DEHN's opinion. In a public talk for a nonmathematical academic audience, DEHN [1928] explained that the continued growth of mathematics requires the continued emergence of new ideas which reduce its complexity. He also expressed the hope that topology would renew its power through such an injection of new ideas. The reader should note the year when this was said. Within the following 10 years, there appeared monographs on topology by ALEXANDROFF and HOPF, by LEFSHETZ, and by SEIFERT and THRELFALL, documenting an upsurge in topological research which has not yet abated.

Chapter I.6

Precursors of Later Developments

In this chapter, we shall briefly describe some papers which motivated or anticipated later developments. We shall have to deal with a great variety of both topics and phenomena. The only feature all of the papers under discussion have in common is a negative one: They did not have both a short-range and a direct effect on the development of combinatorial group theory. Their effect could be short-range but indirect because they were concerned only with finite groups and thus contributed new points of view to group theory which could be transferred with more or less difficulty to the investigation of infinite groups. Such papers are usually not quoted by authors who use their ideas or concepts later on. Other papers mentioned here contain investigations of particular groups which much later attracted a renewed attention. Some papers simply were forgotten and their results and ideas rediscovered by others. We will not always be able to ascertain the reason why a particular result or a particular idea was taken up again in later research. For specific groups or classes of groups which appeared outside of pure group theory, especially in number theory (e.g., the arithmetic theory of quadratic forms), in geometry (discontinuous groups), and in topology (fundamental groups of spaces which are of topological interest), it appears to be a very natural development that they should be the object of later investigations whenever the progress of combinatorial group theory made them possible or whenever new questions involving these groups arose in the fields of their origins. As far as general ideas are concerned, it appears that a steady trend towards generalization and abstraction has played a role in later developments.

The danger of hindsight, always present in historical discussions, is particularly great when we try to trace the development of mathematical ideas. Once we have fully understood a general and abstract principle of construction, we easily recognize its earlier uses in specialized situations and we consider these uses as mildly disguised applications of the general principle. But, in reality, it was a great achievement to extract this principle in a "chemically pure" form, and it made sense to do so only in response to the challenge of new problems. We shall try to illustrate these remarks at the end of this section by analyzing the emergence of the concept of the wreath product. Actually, the emergence of the concept of a group could serve the same purpose, but this is a matter too complex to be dealt with here. In any case, the purpose of these remarks is the following one: We wish to caution against the rash assignment of priorities to particular authors, and we hope that the account of precursors given in this section will not invite unjustified statements of the type of EMMY NOETHER's *Axiom Null: Es steht schon bei Dedekind* (*Axiom Zero: The Theorem Can Be Found Already in the Work of Dedekind*). In the case of EMMY NOETHER, this "axiom" was, of course, an expression of personal modesty since she applied it to her own theorems. But in historical writings, similar axioms are definitely out of place.

We shall try to arrange the papers to be discussed here in collectives which show at least some degree of coherence.

A. Arithmetically Defined Linear Groups in Higher Dimensions

The earliest investigation to be included here appeared in 1866. It was motivated by a problem of two-dimensional topology or, more precisely, of Riemann surfaces. It does not use the words "group" or "matrix." Using present terminology, we may say that CLEBSCH and GORDAN [1866] solved the following problem: The first homology group of a closed orientable two-dimensional manifold of genus g is a free abelian group of rank $2g$. The transition from one homology basis to another is therefore given by an element of the group $GL(2g, \mathbb{Z})$, i.e., of the linear $2g$-dimensional group over the integers. However, topological arguments show that such a transformation must keep a skew-symmetric bilinear form invariant. The question arises whether there are other restrictions to be imposed. The answer is "no," and the proof is given by showing that:

(i)   the symplectic group $Sp(2g, \mathbb{Z})$ is finitely generated and

(ii) generators can be given explicitly, and it can be shown that each generator is in fact induced by a topologically admissible transformation.

CLEBSCH and GORDAN [1866] found such a finite set of generators of $Sp(2g, \mathbb{Z})$ by using purely algebraic methods. Independently, HUA and REINER [1949] again derived a finite system of generators by algebraic methods. Their results are more economical than those of CLEBSCH and

GORDAN, showing, in particular, that at most four generators suffice for all values of g.

Symplectic groups in general, particularly over fields, continued to be of interest as important examples of Lie groups and, in the case of finite fields, as examples of simple nonabelian groups of finite order. The main source of information for the latter aspect is the book by L. E. DICKSON [1901a]. (There the symplectic groups are called "abelian linear groups". The term "symplectic" was introduced by H. WEYL [1939].) However, the groups $Sp(2g,\mathbb{Z})$ did, per se, not attract much attention until SIEGEL [1939 and 1943a] showed that they act as discontinuous groups on symmetric spaces of $g(g+1)/2$ complex dimensions, constructed fundamental regions whose shape showed that the groups are not only finitely generated but also finitely related and constructed automorphic functions in $g(g+1)/2$ complex variables for these groups. But geometry in at least six real dimensions (for the case $g = 2$) is sufficiently difficult to prohibit the derivation of a finite presentation for $Sp(2g,\mathbb{Z})$ in the cases where $g \geq 2$ by the use of geometric methods. The final step in deriving a finite presentation for these groups was taken by BEHR [1975] who used algebraic methods. This information is of interest in the theory of the mapping class groups of two-dimensional closed orientable manifolds. See BIRMAN [1975, especially p. 190].

There are other appearances of $Sp(2g,\mathbb{Z})$ in the literature after 1914 which will be mentioned later in different contexts. Those mentioned so far are the ones connected with the problem of finding a presentation for these groups in the case $g > 1$ (the case $g = 1$ being the same as that of the special linear group in two dimensions). Mathematicians working in the field of analytic functions connected with the theory of Riemann surfaces were always aware of the result found by CLEBSCH and GORDAN in 1866; it appears, after all, in a book on abelian functions. The appearance of these groups in the theory of mapping class groups of two-dimensional manifolds is again perfectly natural, and the timing of this appearance is based on the development of topology and of the theory of moduli of Riemann surfaces. (Some references will be given below.) However, the first appearance of the group $Sp(2g,\mathbb{Z})$ in the work of SIEGEL [1935] announces a surprising event in the development of mathematics during the Twentieth Century. Here the groups appear in the arithmetic theory of positive-definite quadratic forms with integral coefficients in a finite number of variables. The new connections thus established through SIEGEL's work between number theory and important disciplines of analysis involve, as an essential ingredient, the representation of arithmetically defined symplectic groups as discontinuous groups acting on symmetric spaces. (These were studied by E. CARTAN [1936].)

For the linear groups $G_n = GL(n, \mathbb{Z})$ of $n \times n$ matrices with integral entries and determinant ± 1, MINKOWSKI [1905] proved the following theorem:

Let $a_{\nu\mu}, \nu, \mu = 1, \ldots, n$, $a_{\nu\mu} = a_{\mu\nu}$ be $n(n+1)/2$ real numbers which are coefficients of a positive-definite quadratic form in n variables. We consider the $a_{\nu\mu}$ as cartesian coordinates of a point in a part S of $n(n+1)/2$-dimensional euclidean space. Then G_n acts as a discontinuous group on S and has a fundamental region which is a convex cone bounded by finitely many hyperplanes. At any boundary point of this cone other than its apex, there meet only finitely many distinct images of the fundamental region obtained through the action of G_n.

It is obvious from this theorem that G_n has a finite presentation. (Of course, it is trivial that G_n is finitely generated.) MINKOWSKI was not particularly interested in G_n as a group and certainly not interested in finding presentations of groups. His research was motivated by problems from number theory and geometry. However, NIELSEN [1924b] derived a finite presentation of G_n for $n = 3$ without using MINKOWSKI's ideas, although it is extremely likely that NIELSEN was aware of MINKOWSKI's work. Again, it seems that working in a space of six (real) dimensions does not offer the advantages which geometric intuition can provide for problems in two dimensions.

Apart from his general interest in finding presentations for important groups, it seems that NIELSEN had a specific group-theoretical motivation for his paper. We shall discuss its details and its implications in Chapter II.2.

Although MINKOWSKI was not interested in G_n as a group, he was interested in its finite subgroups because they appear as groups of linear transformations which keep a positive-definite quadratic form in n variables and with integral coefficients fixed. In 1887, he showed that the orders of such finite subgroups are divisors of a number $N(n)$ depending only on n and that they cannot be divisible by any prime number exceeding $n + 1$. MINKOWSKI's proof is group-theoretically significant because he proves and uses the fact that a finite subgroup of $SL(n, \mathbb{Z})$ has trivial intersection with any congruence subgroup $\bmod p$ where p is either 4 or an odd prime number. It may be considered as a forerunner of the idea to study a group systematically by studying its finite quotient groups which was systematized by P. HALL when he coined the term "residually finite" (as mentioned by GRUENBERG [1957]). However, it does not appear that MINKOWSKI's work had any direct influence on group theory other than establishing the importance of the groups involved in his research.

Two very general theorems about countable linear groups were proved by A. HURWITZ. The first of these (HURWITZ [1895]) states:

Let R be the ring of algebraic integers in an algebraic number field. Let $SL(n, R)$ be the group of $n \times n$ matrices with the determinant $+1$ and entries from R. Then $SL(n, R)$ is finitely generated.

The main difficulty in proving this theorem arises for $n = 2$. For $n = 3$, the proof is still carried out explicitly by HURWITZ with the remark that the cases $n > 3$ can be dealt with in the same manner. A noteworthy feature of the proof is the use of congruence subgroups.

It appears that HURWITZ' paper has largely been forgotten. We have not been able to locate a reference to it in the extensive current literature on countable linear groups. Of course, it appeared in a nearly inaccessible journal, since most mathematics departments would not subscribe to a periodical which published many papers from nonmathematical disciplines. The publication of HURWITZ' collected papers in 1931 may have come too late, or else their arrangement under the headings of "number theory" and "function theory" may have kept group theorists from looking closely at them.

HURWITZ [1905] also proved the following general result.

Let G be a group of $n \times n$ matrices with determinant $+1$ and entries from the field $\mathbb{C}$ of complex numbers. Assume that G is discrete, that is, assume that there exists no infinite sequence M_r, $r = 1, 2, 3 \ldots$ of distinct matrices M_r in G such that

$$\lim_{r \to \infty} M_r = I,$$

where I is the unit matrix and the limit refers to the n^2 limits of the entries of the matrices. Then G has countably many elements and acts as a discontinuous group on the space of coefficients of the positive-definite Hermitian forms in n variables.

In the same paper, HURWITZ also determined a fundamental region for the Picard group in a space of three real dimensions. Here the Picard group is defined as the unimodular group of 2×2 matrices with entries from the ring of Gaussian integers. This result had been obtained earlier by BIANCHI [1892] and by FRICKE and KLEIN [1897] who also derived a presentation for the Picard group from the construction of the fundamental region.

The papers of HURWITZ are also important as a source of references. He quotes the earlier papers by BIANCHI, BLUMENTHAL, FRICKE, FUBINI,

MINKOWSKI, and STUDY on related topics. Some of these will also be mentioned below.

For the discontinuous groups of n-dimensional euclidean motions, BIEBERBACH, [1911] proved that any such group which has a finite fundamental region contains a free abelian normal subgroup of rank n which consists of translations, and he also showed that for a fixed n there exist only finitely many nonisomorphic groups of this type. The proof of this result was simplified by FROBENIUS [1911], who also proved that even the number of conjugacy classes of such groups within the general affine groups of n dimensions is finite.

Let J_n denote the class of maximal groups of this type (for n fixed). McCARTHY [1970] proved that the groups in J_n can be defined as abstract groups by the property that they have a free abelian subgroup of rank n and that they are "just infinite." This means that the groups themselves are infinite but that all of their proper quotient groups are finite. This is one of the few examples known where a class of geometrically defined groups can also be defined in a simple manner by using only concepts from abstract group theory. (Another example is the characterization of knot groups in more than four dimensions by KERVAIRE [1965].)

B. Arithmetically Defined Linear Groups in Two Dimensions

The bulk of the very extensive literature which falls under the above heading does not belong in a history of combinatorial group theory but in the theory of Riemann surfaces, automorphic functions, and discontinuous groups. FRICKE and KLEIN [1897] and FRICKE [1913] provide good surveys of the results belonging here, and a more recent partial summary may be obtained from MAGNUS [1974a].

Given a set of Moebius transformations

$$ Z \to \frac{\alpha Z + \beta}{\gamma Z + \delta}, \qquad \alpha\delta - \beta\gamma = 1 $$

of the complex variable Z, it is not difficult to show (see FRICKE and KLEIN [1897]) that these transformations generate a group G which acts discontinuously on a part of the complex plane ("Kleinian" groups) or on hyperbolic (non-euclidean) three space if and only if the group G generated by the matrices

$$ \begin{pmatrix} \alpha & \beta \\ \gamma & \delta \end{pmatrix} $$

is discrete in the sense defined earlier in this chapter. In the case where $\alpha, \beta, \gamma, \delta$ are the algebraic integers from a totally real algebraic number field K, BLUMENTHAL [1903 and 1904] showed how one can define a discontinuous action of G on a space of n complex dimensions where n is the degree of K over the rationals. G is called "Hilbert's modular group." We shall not be concerned here with this case, and we shall not even discuss the majority of cases where an arithmetically defined group G acts discontinuously on part of the complex plane or on hyperbolic three space. Instead, we shall merely discuss a few special cases where the groups involved appear again later in purely group theoretical investigations and not merely because of their discontinuous action on certain spaces.

We shall consider the following groups: first, the elliptic modular group M, consisting of the Moebius transformation in which $\alpha, \beta, \gamma, \delta \in \mathbb{Z}$; next, the Picard group P, where $\alpha, \beta, \gamma, \delta$ are Gaussian integers; and finally, the Bianchi groups B_D, where $\alpha, \beta, \gamma, \delta$ are algebraic integers of the imaginary quadratic number field $\sqrt{-D}$, where D is a positive integer. Of course, in all cases $\alpha\delta - \beta\gamma = 1$. Note that $P = B_1$.

The modular group M and its subgroups have been dealt with extensively by KLEIN and FRICKE [1890 and 1892]. It can be generated by two Moebius transformations a and b defined by the matrices

$$\begin{pmatrix} 0 & -1 \\ 1 & 0 \end{pmatrix} \qquad \begin{pmatrix} 1, & -1 \\ 1 & 0 \end{pmatrix}$$

and can be presented by the defining relations

$$a^2 = b^3 = 1.$$

These facts follow easily from the construction of the fundamental region of M in the upper half of the Z-plane on which M acts as a discontinuous group, and it is difficult to say who first constructed this fundamental region. The construction is implicit in the reduction theory of quadratic binary forms and in this sense was known to GAUSS who, of course, cannot be credited with the presentation of M. On the other hand, it is obvious that it was known to DYCK [1882].

The subgroups of M, and in particular the normal subgroups of finite index, are of great interest for reasons which belong to the theory of algebraic equations, of algebraic number theory, and of the theory of automorphic functions. They are fully explained by KLEIN and FRICKE [1890 and 1892] and FRICKE [1926] and cannot be discussed here. Of particular interest are the so-called *principal congruence subgroups* M_n of *level* n consisting of those Moebius transformations whose matrices are congruent to $\pm I \pmod{n}$, where I is the unit matrix. KLEIN [1880] observed, and FRICKE [1886] and PICK [1886] proved simultaneously and independently, that M has normal subgroups and even normal subgroups of finite index which do not contain any subgroup M_n. The proofs of FRICKE and of PICK use elegant number-theoretical arguments. There exists an essentially group-theoretical proof which is based on the Jordan–Hoelder theorem and on the fact that in the composition series of M/M_n, only cyclic groups and groups $PSL(2, p)$ can occur. This proof (given explicitly in MAGNUS [1974a] with additional information) may very well go back to the time before 1914. However, we have been unable to locate it in the literature.

MENNICKE [1965] and, in a larger context, BASS, MILNOR, and SERRE [1967] proved that for $n \geq 3$, the groups $SL(n, \mathbb{Z})$ and large classes of other linear groups have the so-called "congruence subgroup property" which states that, apart from subgroups of the center, every normal subgroup contains a congruence subgroup. An immediate consequence of this result is the following purely group-theoretical theorem:

There exist infinitely many explicitly constructible finitely presented infinite groups in which every normal subgroup is of finite index and where the intersection of the normal subgroups is the unit element. In addition, these groups do not contain an abelian normal subgroup.

Until now, such a group can not be constructed with purely group-theoretical methods. However, there exist general theorems about their structure (see WILSON [1971]).

The group M is probably the most extensively studied infinite group with a finite presentation. We shall skip references to most of the literature which describes presentations for subgroups of M. Since M is the free product of two cyclic groups of orders 2 and 3, respectively, a theorem due to KUROSH [1934] describes the abstract structure of all possible subgroups completely, and it is not the abstract but the concrete number theoretically defined structure of the subgroups which has been studied. However, we mention a paper by NEUMANN [1933] which describes the construction of a set S of subgroups of M characterized by the fact that these subgroups contain the generator a and are such that in the matrices defining their elements, every pair of coprime integers appears exactly once as the first column of such a matrix. Remarkably, the cardinality of S is that of the continuum. Finally, we mention a result of LEVIN [1968], who proved a conjecture of B. H. NEUMANN according to which every countable group is isomorphic to a subgroup of a quotient group of M. A good deal of the development of combinatorial group theory can be illustrated by its effects on our knowledge of the modular group M. The result of NEUMANN [1933] requires for its proof an algorithm for the construction of subgroups of a group given by a presentation, the so-called Reidemeister–Schreier method; see SCHREIER [1927a]. The characterization of the subgroups of M as free products of cyclic groups of order 2, 3, or ∞ requires the fundamental subgroup

theorem of KUROSCH [1934]. And the tools needed to prove the result due to
LEVIN [1968] include several of the most powerful methods developed in
combinatorial group theory. However, it would be too much of a digression
to list them here.

The groups B_D all have the obvious property of being discrete and the
slightly less obvious property of not being discontinuous in any part
of the complex Z-plane. BIANCHI [1892, 1893a, and 1893b] and HUMBERT
[1915–1920] gave generators for these groups and carried out, or at least
indicated, the construction of a fundamental region for them in hyperbolic
three space. The construction is particularly explicit in BIANCHI [1892] for
$D = 1$, 2, 3, 5, 6, 7, 10, 11, 13, 15, 19 and in HUMBERT [1915–1920] for
$D = 21$.

The group $P = B_1$ had been studied earlier by PICARD [1896], who had
found generators and a fundamental region in hyperbolic three space for it,
and FRICKE and KLEIN [1897] devoted a whole chapter of their book to it
and used their construction of the fundamental region to derive a presentation for P. However, the difficulties of deriving a presentation of a discontinuous group from the shape of its fundamental region are already considerable in three (real) dimensions, and it was not until nearly a century later
that SWAN [1971] found an effective method of obtaining presentations for
the groups B_D, giving explicit results for $D = 1$, 2, 3, 5, 6, 7, 11, 15, and 19.
SWAN's work is based in part not only on the papers by BIANCHI and
HUMBERT but also on a topological result due to MACBEATH [1964]. SWAN
[1971] also proved general theorems about the groups B_D for sufficiently
large D, some of them connected with number-theoretical properties of the
ring of algebraic integers in the number field generated by $\sqrt{-D}$. For
$D = 1, 2, 3, 7, 11$, FINE [1974] gave a presentation for B_D which is derived
purely algebraically and based on the work of COHN [1968]. FINE's paper
also gives an insight into the structure of B_D for the special values of D
mentioned by showing that they have subgroups of finite index which can
be constructed as free products with amalgamations of groups with a
well-known simple structure. For $D = 1$, this was done earlier by WALDINGER
[1965] and DRILLICK [1971].

PICARD's group P has been given a little more attention than the other
groups B_D. This may be due to the fact that FRICKE and KLEIN [1897]
derived a simple presentation for P and also gave a brief history of the
research involving the group up to that time and also pointed out the
importance of P for the arithmetic theory of binary quadratic and Hermitian forms with coefficients from the ring of Gaussian integers which had
been started by DIRICHLET and HERMITE. For references to these matters,
see FRICKE and Klein [1897, pp. 91–93]. Later investigations include a
detailed study by SANSONE [1923], who also gave a new presentation of P,
and the construction of infinitely many normal subgroups of P which are of

finite index and do not contain any congruence subgroup by DRILLICK [1971].

C. Geometric Constructions. Fuchsian Groups

The finitely generated discrete groups of Moebius transformations which map the interior of a circle in the complex plan onto itself are called fuchsian groups. They do not exhaust the class of all of the groups normally covered by this name, but for our present purposes it is convenient to use the term "fuchsian" for these groups only. If we supplement this class by the discrete groups of planar euclidean motions and by the finite groups of Moebius transformations, the resulting class of groups was shown by FRICKE and KLEIN [1897, pp. 186–187] to coincide with the class of finitely generated groups which have a presentation of the following type.

Generators: c_ρ $(\rho = 1,\ldots,r)d_\sigma$, $(\sigma = 1,\ldots,s)$, a_ν, b_ν, $(\nu = 1,\ldots,g)$
Defining Relations:

$$d_\sigma^{e_\sigma} = 1(e_\sigma \geqq 2), \text{ and } c_1 \cdots c_r d_1 \cdots d_s a_1 b_1 a_1^{-1} b_1^{-1} \cdots a_g b_g a_g^{-1} b_g^{-1} = 1.$$

The possibilities that one or two of the sets of c_ρ or of d_σ or a_ν, b_ν are empty are not excluded.

The method used by FRICKE and KLEIN to prove this theorem is geometric. Their result is based on the construction of fundamental regions in the complex plane which have shapes of a particular type. The proofs are not at all easy and, in part, may not be considered fully rigorous by today's standards, but the results are undoubtedly correct, having been confirmed later with the help of more powerful tools.

It appears that FRICKE (who is the author responsible for the larger part of the contents of the work by FRICKE and KLEIN [1897]) considered the presentations of fuchsian groups as an important way of defining them in a concise and purely group-theoretical manner but found very few opportunities for using them in other investigations.

However, FRICKE's presentations for the members of a class of groups with geometric as well as topological significance (to be explained below) have been used in a great many later papers dedicated to problems of combinatorial group theory. Among other things, they have been used as test cases or as examples for the applicability of new theorems. Since the torsion-free fuchsian groups are also fundamental groups of two-dimensional orientable manifolds, the development of topology has also led to the emergence of new questions concerning these groups. Probably the most important of such questions was the one raised by H. HOPF in 1932 at the International Congress of Mathematicians in Zuerich. HOPF asked whether such a group could be isomorphic to a proper quotient group of itself. Obviously, this is a question which can be asked about every group given by a presentation and it is a highly nontrivial question if the group is finitely generated but infinite. Strangely enough, it seems that HOPF was the first to call the attention of group theorists to this question.

Describing in detail the occurrence of fuchsian groups in later work on combinatorial group theory would involve lengthy discourses which have to be placed in the account of the history of a later period. However, here we have to mention the Ph.D. thesis of H. GIESEKING [1912] which had been instigated by MAX DEHN. In the first part of this paper, the geometric construction of a fundamental region for the group Φ_2 with generators $a, b, c\, d$, and defining relation

$$aba^{-1}b^{-1}cdc^{-1}d^{-1} = 1$$

is carried out explicitly by constructing a regular octagon in the non-euclidean plane (represented by the interior of the unit circle) with center at the center of the unit circle and with angles $\pi/4$. The generators of Φ_2 are then represented explicitly by 2×2 matrices with determinant $+1$ and with entries from an algebraic number field. They define the Moebius transformations which map the unit circle onto itself and move the octagon in the proper manner onto an adjacent one. The resulting matrix representation for Φ_2 is then used for group-theoretical discussions. The faithfulness of the representation is not directly demonstrated but derived from the fact that it reflects the geometric construction.

The second part of the thesis of GIESEKING [1912] contains the only purely geometric construction of a fundamental region of a group in non-euclidean three space which was carried out before 1918. The construction leads to a presentation of the group. It is modelled after the corresponding construction in the non-euclidean plane where the reflections in the sides of a triangle with angles zero produce a group with three generators and with relations which state that the squares of the generators are the identity. In non-euclidean three space, the triangle is replaced by a tetrahedron; the construction and the presentation of the resulting group are described briefly in MAGNUS [1974a, pp. 153–155]. The original paper by GIESEKING is nearly inaccessible.

D. Braid Groups and Mapping Class Groups

HURWITZ [1891], in a paper on the theory of Riemann surfaces, discovered and investigated a class of groups which later—with a different definition—became known as "braid groups." This involves a special—although important—class of groups. The connection between the paper by HURWITZ [1891] and later developments has been described in detail by MAGNUS [1974b]. We shall briefly come back to this topic in Chapter II.10. However, we wish to point out that we have here at hand a rather extreme

and, as we believe, unusual case of neglect of an important paper. HURWITZ'
paper may very well be interpreted as giving a first approach to the idea of a
fundamental group of a space of arbitrarily many dimensions. In fact, one
may say that HURWITZ' paper is, in part, an illustration of the theorem that,
in a fiber space, the fundamental group of the base space acts as a group of
self-mappings on the fiber. Of course, HURWITZ would not have had at his
disposal the general concepts appearing in such a statement since these
emerged much later after a long development in topology. However,
HURWITZ' paper is written with great clarity. The topological arguments are
in a shape which may be considered as rigorous even though one might
express them differently today. In addition, HURWITZ was a mathematician
with a great name, and he published his paper in one of the most widely
read journals of his time. Nevertheless, it seems that his share in the
discovery of braid groups is mentioned for the first time in MAGNUS
[1974b]. Other aspects of HURWITZ' paper have been quoted widely and
frequently, but in different contexts.

HURWITZ [1891] also introduced some ideas which may be considered as
the beginning of the theory of mapping class groups of two-dimensional
manifolds. But the bulk of contributions to this topic (up to 1914) is due to
FRICKE and KLEIN [1897, pp. 285–447 and 1912, pp. 286–439]. Only a
small part of it belongs in a history of combinatorial group theory, and we
refer summarily to the literature and to later developments by quoting
MAGNUS [1974b] and BIRMAN [1975].

The Ph.D. thesis of J. NIELSEN [1913] consists of two parts. The first
three sections deal with a topic suggested to NIELSEN by LANDSBERG
involving differential geometry. The last section deals with the problem of
determining the minimal number of fixed points of a one-to-one topological
self-mapping of a torus. It arose from a question raised by M. DEHN who
became NIELSEN's thesis advisor after the death of LANDSBERG. The homogeneous modular group $GL(2, \mathbb{Z})$ plays a role in this paper. However, its
main importance is derived from the fact that NIELSEN pursued the fixed-
point problem for topological self-mappings of two-dimensional manifolds
in many later publications, some of which also contained important group-
theoretical contributions to the theory of mapping class groups. For references and a few details see Chapter II.10.

E. Differential Equations, Linear Groups,
and Lie Groups

SCHLESINGER [1909] gave a survey of the theory of linear differential
equations with a list of 1,742 references to papers published in the years
1865 to 1907. Schelsinger quotes many papers on linear groups (over fields

of characteristic zero). Generally speaking, these papers have contributed to later developments of combinatorial group theory only in the same sense in which this is true for the theory of abstract finite groups, namely, by raising the level of sophistication and by introducing new aspects into the investigation of groups. However, there are a few papers which may be singled out as precursors of later developments in combinatorial group theory, although some of them had an effect only after new and powerful tools from algebra and algebraic geometry had become available.

We shall begin our account with a report on an investigation which goes back to POINCARÉ [1884], who posed the following problems.

1. Given a linear differential equation with algebraic coefficients, determine its group.
2. Given a linear second-order differential equation which depends on certain arbitrary parameters, dispose of these parameters in such a manner that the group of the equation is fuchsian.

The second problem shows the origin of the questions raised: they come from the theory of automorphic functions. POINCARÉ himself makes that clear by quoting, at the beginning, two of his previous papers published in the same journal whose titles alone confirm our statement (see POINCARÉ [1882 and 1883]).

The group mentioned in the first problem stated by POINCARÉ is the monodromy group of the linear (homogeneous) differential equation. After choosing a basis of n linearly independent solutions of the differential equation (n being its order) in the neighborhood of a regular point, one obtains a certain number of linear substitutions for these solutions by continuing them analytically along k simple closed loops around the k singular points of the equation. By a simple transformation of the equation, one can arrange the matrices of these substitutions to have determinant $+1$. These matrices then generate the monodromy group M. A change of basis has the effect of conjugating all matrices by a fixed nonsingular matrix. POINCARÉ then deals with the following question (which is quite independent of the theory of differential equations): Which data are needed to determine the conjugacy class of the group M within the special linear group $SL(n, \mathbb{C})$ of $n \times n$ matrices with determinant $+1$ and entries from the field $\mathbb{C}$ of complex numbers? These data are called invariants, and POINCARÉ shows that the coefficients of the characteristic equations of the generators of M and of finitely many of their products suffice "in general" to determine the conjugacy class of M in $SL(n, \mathbb{C})$ uniquely. (There are obvious exceptions, e.g., a group generated by upper triangular matrices.)

For the case $n = 2$, H. VOGT [1889] proved sharper results. "In general," $3k - 3$ invariants of a k-generator subgroup of $SL(2, \mathbb{C})$ are needed to determine not one but at most 2^{k-2} conjugacy classes in $SL(2, \mathbb{C})$. One also may interpret some of VOGT's results as the first determination of a system of generators for the automorphism group of a free group. Subsequently, FRICKE and KLEIN [1897 and 1911] developed the theory further to a study of the mapping class groups (see BIRMAN [1975]) of bounded or unbounded two-dimensional orientable manifolds, and much later, HOROWITZ [1972 and 1975] gave a systematic algebraic foundation with purely group-theoretical applications to some of the results of FRICKE and KLEIN [1897 and 1912]. POINCARÉ's original investigation also has been taken up again and both vastly expanded and systematized, at least in its algebraic aspects, by PROCESI [1975], LERON [1976], and RAZMYSLOV [1974].

We can not document much of a noteworthy direct influence of this work of POINCARÉ, VOGT, FRICKE, and many others on purely group-theoretical research in later times. But it is probably safe to say that the interest of group theorists in finitely generated linear groups has in part been maintained by their importance for non-group-theoretical problems. The survey by WEHRFRITZ [1973] shows how much work has been done here since the beginning of the Twentieth Century. These groups have some surprisingly simple properties, particularly if the entries of the generating matrices belong to a field of characteristic zero. In this case (but not only in this case), MAL'CEV [1940] showed that the groups are residually finite. Later, SELBERG [1960] proved that the groups contain a torsion-free normal subgroup of finite index. And TITS [1972] proved that these groups either contain a free subgroup of rank 2 (and, therefore, of infinite rank) or they are finite extensions of solvable groups. It should be noted that the second alternative includes the groups of upper triangular matrices for which POINCARÉ's theorem about the characterization of their conjugacy class by a finite number of invariants is not valid.

Of the three theorems mentioned here, only SELBERG's theorem would have been at least conceptually within easy reach of mathematicians before 1914. The term "residual finite" was introduced by P. HALL in 1955. Looking backwards, one may feel that the concept arises naturally from the well-known properties of the quotient groups of congruence subgroups in matrix groups with entries from algebraic number fields. MAL'CEV [1940] certainly was fully aware of this fact, although he did not coin the term. But fruitful abstractions look natural only in hindsight.

Finally, TITS' theorem could not have been formulated easily before 1914, in spite of the fact that free groups were well known. But they were known mainly as a sort of "most general" group. Their appearance as subgroups in practically all fuchsian groups became known only after REIDEMEISTER [1926] and SCHREIER [1927] had developed a method of computing a presentation for a subgroup from that of the whole group. As for the methods used by TITS, e.g., the Zariski topology (see e.g., WEHRFRITZ [1973]), they were completely out of reach before 1914.

There exists another aspect of the theory of ordinary differential equations which eventually, via a tremendous detour, led to the development of important tools for the investigation of purely group-theoretical problems. The theory we are referring to was known originally as the Picard–Vessiot theory and was later called the theory of algebraic groups. For a long time, one of its motivating purposes was to find the conditions under which the solutions of ordinary linear differential equations can be expressed in a closed form with the help of quadratures and certain functions which included, of course, the algebraic functions and coefficients. From its very beginning, the theory had been conceived as a generalization of the Galois theory of algebraic equations. It led to the theory of groups of matrices whose entries satisfy certain algebraic equations (e.g., those characteristic for orthogonal matrices) and eventually made it possible to prove, for instance, the following theorem due to PICKEL [1971], which is purely group theoretical and shows no connection whatsoever with differential equations:

Let P be a finitely generated nilpotent group, that is, a group in which only finitely many terms of the lower central series are different from the unit element. Consider the set S of all distinct finite quotient groups of P. Then there exist only finitely many nilpotent groups which have the same set S of finite quotient groups as P.

So far, it is not possible to prove this theorem without using the theory of algebraic groups.

We have quoted this result as an illustration of the coherence of mathematics. We cannot try to give a full account of the development which led to it or do justice to the many mathematicians who contributed to it. We merely quote the original papers by PICARD [1896] and VESSIOT [1892], an early and very lucid systematic foundation of the theory by LOEWY [1908], an important survey by RITT [1950], and the lecture notes of HUMPHREYS [1975] which contain references to later papers.

LIE's theory of transformation groups or "continuous groups," as they were called for a long time, was originally connected with the theory of partial differential equations. One of the last important papers during the period considered and showing this connection clearly is probably the investigation by EMMY NOETHER [1918], who derived the "intermediate integrals" of a system of partial differential equations from the Lie group which keeps the system invariant.

There is a clearly visible, although not very well documented, early influence of the theory of Lie groups on the theory of finite groups. The construction of simple groups of finite order by L. E. DICKSON [1901a] in his standard work on (finite) linear groups and even more so in his subsequent papers (DICKSON [1901b and 1905]) obviously runs parallel to the construction of simple Lie groups although, of course, the methods are not the same. The characterization of finite groups which are direct products of groups of prime power order by BURNSIDE [1911, p. 166] shows an unmistakable transfer of the concept of a nilpotent group from Lie groups to finite groups. Today (1980) one speaks of "finite simple groups of Lie type."

We know through a letter from P. HALL that he became interested in group theory by reading the book by BURNSIDE [1911] on the theory of groups of finite order. This fact establishes a link between BURNSIDE's work on groups of prime power order and the paper by P. HALL [1933] on the same topic. The title of HALL's paper, *A contribution to the theory of groups of prime power order*, does not reveal the fact that it also represents a milestone in the history of combinatorial group theory. This is due to the section on "commutator calculus" in the paper in which HALL systematically investigates the rather complicated relation between the associative composition of two group elements and the composition defined by forming their commutator. The result is the first thorough analysis of the groups of the lower central series (a term coined by HALL) of free groups and their relation to other fully invariant subgroups of free groups.

What emerged eventually, via a rather tortuous detour, was the fact that the lower central series of both free groups and of many other groups defines a Lie algebra and that one is able to see the root of this connection in papers by CAMPBELL [1898], BAKER [1904], and HAUSDORFF [1906] from the turn of the century. These papers were directly motivated by the theory of Lie groups.

The relationship between lower central series and associated Lie algebras is presented in detail in Chapter 5 of MAGNUS, KARRASS, and SOLITAR [1966]. There a proof of the so-called "Campbell–Baker–Hausdorff formula" is also given. For the early history of this formula, see W. SCHMID [1982]. In combinatorial group theory, it was used for the first time in the Ph.D. thesis of ADELSBERGER [1930], where its first two terms were utilized for the investigation of nilpotent groups of class 2, (i.e., of groups for which the third term of the lower central series is trivial). The thesis was based on an idea of REIDEMEISTER [1927] (the thesis adviser). However, no reference to the original paper by HAUSDORFF [1906] was given, although REIDEMEISTER told MAGNUS in 1936 that this paper had given him the idea for the topic of the thesis. Subsequently, MAGNUS, who recognized the connection of the formula to his own work, utilized the full result in 1950. There are later papers using the formula, but its usefulness has so far been rather limited, except for the fact that it strongly suggests connections between Lie algebras and groups which, however, must then be proved without using the formula. We shall discuss the work of P. HALL in Chapter II.7.

F. Finite Groups

After the publication of CAMILLE JORDAN's *Traité des substitutions* in 1870, the theory of finite groups developed rapidly. Numerous textbooks in English, French, German, Italian, and Russian appeared, some of them dedicated entirely to group theory, others containing chapters on it as an important part of algebra, specifically of Galois theory. The theory of representations of finite groups as groups of finite matrices, developed by FROBENIUS, BURNSIDE, I. SCHUR, and others, introduced new aspects into the theory. A large number of contributions, some of them of outstanding importance, came from the United States, where interest in pure mathematics was growing fast and where B. S. EASTON [1902] published a bibliography of the field.

In view of the great and widely disseminated amount of knowledge about finite groups existing after 1914, it is, in many cases, impossible to trace later developments in combinatorial group theory to earlier ones in the theory of finite groups by quoting particular papers. For instance, the search for finitely presented simple infinite groups was, after 1920, a perfectly natural question. That it has been successful only lately (R. J. THOMPSON, G. HIGMAN; see HIGMAN [1974]) is undoubtedly due to the glaring difficulty of the problem which reveals itself almost immediately. Other concepts from the theory of finite groups which invite generalizations in a natural manner are those of solvability (defined by the termination of the derived series) and of nilpotency (already discussed above). Here the difficulties which arise, even for finitely generated groups, are again considerable except for the case of abelian groups, but progress has been more or less steady because it is possible to find "reasonable" additional restrictions which lead to manageable problems. In the case of the Jordan–Hoelder theorem, which ceases to be valid even for infinite cyclic groups, HIRSCH [1938] showed that it is nevertheless possible to salvage a weakened form of the theorem by restricting oneself to the class of solvable groups with maximum condition for subgroups. These groups are called "polycyclic" because they have a finite normal series with cyclic quotient groups. These need not be pairwise isomorphic for different normal series, but the number of infinite quotient groups is always the same. Polycyclic groups always have a finite presentation. For this and related topics, see ROBINSON [1972] and Chapter II.10.

To describe all results arising from the theory of finite groups in the manner indicated above would involve a survey of a large part of combinatorial group theory up to the present time. We cannot do this here, and we even have to abstain from mentioning the transfer of other concepts from finite to infinite groups which, like that of the Frattini subgroup (i.e., the subgroup generated by elements which cannot appear in any minimal set of generators) have a distinctly "combinatorial" flavor.

Some methods and constructions which originated in the theory of group representations have been utilized for combinatorial group theory. Here the transfer is less "natural" than in the cases mentioned above, and we shall describe the situation in greater detail.

A first example is the method of constructing all representations of a (finite) group as a transitive permutation group by considering the action of the group elements on the right cosets of a subgroup. We shall not try to trace the origins of this method but merely mention that it appears, with all details and consequences, in FROBENIUS [1895]. It was then used by SCHREIER [1927] for a very elegant solution of the word problem of free groups, i.e., for a proof of the fact that different freely reduced words in the free generators represent different group elements. (SCHREIER calls this fact the "existence of free groups.") Naturally, SCHREIER does not quote FROBENIUS or anybody else. The method itself was then common knowledge.

A more complicated case is presented by the discovery of the wreath product $A \wr B$ of two groups A and B by KALOUJNINE and KRASNER [1918] (who called it *le produit général*). Since $A \wr B$ can indeed be derived from the regular representation of A as a group of permutation matrices in which the nonzero entries are replaced by arbitrary elements of B, one may say that the construction goes back to earlier authors, e.g., FROBENIUS [1898 and 1899]. KERBER [1971] gives other references to potential precursors of the construction. Here, however, one has to note two things. First of all, KALOUJNINE and KRASNER give an abstract definition of $A \wr B$. But more important is the fact that they proved the following theorem:

Every group G which contains a normal subgroup isomorphic with B such that $G/B \simeq A$ is isomorphic with a subgroup of $A \wr B$.

This result makes the wreath product a powerful tool in the theory of varieties of groups. For details, see H. NEUMANN [1967].

Probably the most remarkable case of a tool of combinatorial group theory which has its roots in the early (before 1914) theory of group representations is that of the SCHUR multiplicator. The problem from which SCHUR [1904] starts is the following one: Let H be a finite group of fractional linear substitutions with $n \times n$ matrices $a, b, \ldots$. Then these matrices $a, b, \ldots$ themselves form a group G which contains an abelian subgroup A in its center such that $G/A \simeq H$. This subgroup A is not uniquely determined since the matrices $a, b, \ldots$ can be replaced by multiples $\lambda a, \mu b, \ldots$ (where $\lambda, \mu, \ldots$ are complex numbers $\neq 0$) without changing H. The problem is to find subgroups A of minimal order. SCHUR then shows that the solution of this problem can be reduced to the construction of a certain abelian group M which he calls the *multiplicator* (or *multiplier*) of the abstract group H and which can be defined by the following properties: There exist groups K (called *Darstellungsgruppen*) associated with H which have the following properties: K contains a subgroup M in its center such that $K/M \simeq H$ and such that M is contained in the commutator subgroup of K and is of minimal order. The group M is the same for all groups K.

SCHUR [1904] showed that one can construct M directly from the multiplication table of H. It can therefore, at least in principle, be constructed from a presentation of M.

It turned out that the multiplier M of H is the second cohomology group of H with coefficients in the group $\mathbb{C}^*$ of nonzero complex numbers, as defined by EILENBERG and MACLANE [1949]. This becomes immediately clear if one reads SCHUR's paper after reading the paper by EILENBERG and MACLANE, who, however, were not aware of this connection in 1949. It is acknowledged in a footnote on p. 137 in MACLANE [1963].

It took some time before the cohomology theory of groups became useful for combinatorial group theory. Today, this is definitely the case, and, in particular, the Schur multiplier has been used by BAUMSLAG [1976a] to prove that certain groups are infinitely related by observing that a group with an infinitely generated Schur multiplier must be infinitely related. (The converse is not true; see BAUMSLAG [1976b].)

We briefly mention that FRICKE and KLEIN [1897, pp. 200–209] treated a very special case of SCHUR's problem for (infinite) fuchsian groups.

Turning now to papers in the theory of abstract groups which much later led to important developments, we must undoubtedly assign first place to a problem raised by BURNSIDE [1902]. Let $B(n, e)$ denote the group on n generators which is defined by the defining relations $x^e = 1$, where n and e are fixed and x runs through all words in the generators. BURNSIDE asked: For which values of n and e is $B(n, e)$ finite? This is trivially true for all finite n and $e = 1$ or $e = 2$. BURNSIDE proved that B is also finite if $e = 3$ and shortly afterwards DE SÉGUIER [1904, p. 72] proved the same for $n = 2$ and $e = 4$. The next four papers on the subject appeared in 1933, 1940, and 1950, written, respectively, by LEVI and VAN DER WAERDEN, SANOV, MAGNUS, and MEIER-WUNDERLI. A survey of the history of the problem until 1960 may be found in MAGNUS, KARRASS, and SOLITAR [1966, pp. 379–393]. Although there still remain many open questions, at least the most difficult part of BURNSIDE's problem has been solved: ADJAN and P. S. NOVIKOV (see ADJAN [1975]) proved that $B(n, e)$ is infinite for all $n \geq 2$ and odd $e \geq 665$. This paper is possibly the most difficult paper to read that has ever been written on mathematics. This is not solely due to its length (335 pages). There are other very long papers, even in group theory, e.g., the proof by FEIT and THOMPSON [1963] that groups of odd order are solvable. But at least a reader familiar with the theory of finite groups and their representations will encounter familiar terms and the use of familiar theorems throughout their paper. Such is not the case with the paper by ADJAN. One has to read it word for word from the very beginning. It is based on extremely sophisticated combinatorial arguments.

A comparison of the influence of BURNSIDE's problem on combinatorial group theory with the influence of Fermat's last theorem on the development of algebraic number theory suggests itself very strongly. Both problems are of a rather special nature. In the case of BURNSIDE's problem, the author himself must have felt this since the result of his 1902 paper appears only as an exercise on p. 143 in his textbook, BURNSIDE [1911]. But in spite of their special nature, both problems have fascinated generations of mathematicians and have never been forgotten. In the case of BURNSIDE's problem, this is not easily documented by references to published papers, but even before the appearance of the paper by LEVI and VAN DER WAERDEN in 1933, DEHN quoted it in colloquium lectures. Finally, both problems have acted as stimulants for the development or the refinement and the testing of new methods. In the case of FERMAT's last theorem, this is well known. In the case of BURNSIDE's problem, the survey of its history until 1960 (quoted above) gives at least an indication. It should be added that the solution of BURNSIDE's problem has enabled ADJAN and NOVIKOV (see ADJAN [1975]) to produce particularly striking examples of phenomena in the theory of infinite groups which contradict all the expectations we may have developed on the basis of our experience with groups that somehow behave like finite groups. In analysis and topology, we have been accustomed for a long time to surprises coming from the infinite. We note here that the cardinality of the continuum is not needed to produce such surprises. The countable cardinality of finite words in finitely many symbols together with simple transformation rules suffices for this purpose. For more details on BURNSIDE's problem, see Chapter II.7.

In the same year as BURNSIDE's paper, there appeared a paper by HOYER [1902] of about the same length and the same level of mathematical difficulty. But its fate was the opposite of BURNSIDE's paper, namely, nearly complete oblivion. (Our attention has been called to it through a communication from W. GASCHUETZ to G. BAUMSLAG.) And yet it anticipates both in its method and in its result a much quoted formula found by SCHREIER [1927] which is fundamental for an important technique in combinatorial group theory.

In current language, the contents of the paper may be described as follows: Let Γ_ν be a free semigroup on ν generators with right-hand but not left-hand cancellation. If we postulate relations between the generators (i.e., equations of the form $W_1 = W_2$, where W_1, W_2 are words in the generators and where $W_1 = W_2$ implies $W_1 W = W_2 W$ for all words and vice versa), the elements of Γ_ν are classes of elements equal to each other. Suppose we introduce so many relations that the number of equivalence classes becomes a finite number n. Then these equivalence classes form a group G. (This need not be true if n is infinite.) HOYER proves the theorem:

We need at least $n\nu - (n - 1)$ relations to obtain a finite number of equivalence classes. On the other hand, for any finite values of ν and n, we can always find

$nv - (n - 1)$ *relations which change* Γ_ν *into a given finite group of order n with* ν *generators.*

HOYER then observes that, in a group, both cancellation laws hold although in Γ_ν the left-hand cancellation law cannot be derived from the right-hand cancellation law. He also notes that a finite group of order n with ν generators can, in general, be defined by fewer than $nv - (n - 1)$ relations if we assume both cancellation laws.

The number $nv - (n - 1)$ is, of course, exactly the number of free generators of a subgroup of index n in a free group on ν free generators which had been found by SCHREIER [1927]. It is remarkable that the proofs by HOYER and SCHREIER also use exactly the same method: HOYER represents the elements of G by words (i.e., by elements of Γ_ν) of minimal length. This is exactly the definition of a minimal Schreier system of coset representatives of a subgroup of a free group. It should also be noted that HOYER not only introduces the concept of a semigroup before DE SÉGUIER [1904], who is usually credited with the introduction of this concept, but that, in addition, he also recognizes the role of the cancellation laws (which DE SÉGUIER assumes to be valid) and the independence of the right-hand and left-hand cancellation laws.

It is almost certain that SCHREIER was not aware of HOYER's paper. Its title, *On the definition and the treatment of transitive groups*, certainly does not reveal its contents. There appeared numerous papers on transitive permutation groups at the time HOYER's paper was published, and these had nothing to do with arguments of the type HOYER used. So it is understandable that people interested in the emerging theory of combinatorial aspects of group theory overlooked HOYER's paper then and later on. On the other hand, HOYER was, for a while at least, active in research. Papers by him on algebra and group theory appeared from 1897 to 1906. Apparently, he did not try to find any further applications of his approach to the theory of groups that are defined by generators and relations. The only explanation we can give for this fact is a rather vague one. It was simply too early. After all, nobody investigated subgroups of free groups systematically before NIELSEN [1921]. HOYER's work was rediscovered by GASCHUETZ [1956], who refers to it.

The last influence from the pre-1914 theory of finite groups on combinatorial group theory to be mentioned here is not fully documented but based on what we consider a reasonable guess. Together with REMAK [1911], O. SCHMIDT [1913] was the first to generalize the theorem of the unique decomposition of groups into a direct product of directly indecomposable factors from finite abelian groups to the class of all finite groups.

Since SCHMIDT [1916] emphasized the importance of including, as far as possible, infinite groups into a general theory of abstract groups, and since

KUROSH attended the seminar of SCHMIDT in 1930, we conjecture that KUROSH's paper of 1934 was motivated by an attempt to transfer the theory of direct products of groups to a theory of the newly discovered free product of groups for which KUROSH gives credit to SCHREIER [1927a]. The result is, among others, the proof of the Kurosh subgroup theorem which may be considered as the first general structure theorem of combinatorial group theory. Rather surprisingly, the theory of decomposition of groups into freely indecomposable factors turns out to be considerably simpler than the theory of the decomposition into directly indecomposable factors. The details of the theory of free products will be described in Chapter II.4.

Chapter I.7

Summary

In this chapter, we shall try to describe the emergence of combinatorial group theory in the period up to 1918 as a new branch of group theory, using a minimum of technical language and avoiding minutiae as well as such documentary quotations which have to be interpreted to be understood today.

Group theory started as the theory of groups of transformations, i.e., of one-to-one and onto self-mappings of a mathematical object. If the object is a finite set, the group appears as a permutation group. In this form, group theory became a prominent tool in the theory of algebraic equations, and the work of CAMILLE JORDAN on "substitutions and algebraic equations" was the first lasting monument of these ideas. It appeared in 1870. Its topics are finite groups, and the simplest prominent feature of a finite group is the factorization of its order into prime numbers. JORDAN's book not only presented the knowledge of finite groups and their role in the theory of equations which was available at that time in a well organized fashion; it also contained new results, and the appearance of his work has, according to contemporary sources, greatly accelerated the development of the theory of finite groups.

Infinite transformation groups became a topic of extensive research at about the same time. But this development was initiated not by a fundamental work but by a manifesto. In 1872, FELIX KLEIN gave the public lecture (*Antrittsvorlesung*) at the University of Erlangen which was mandatory for newly appointed professors.

Like BERNHARD RIEMANN in 1854 in his lecture initiating him as a faculty member at the University of Goettingen, KLEIN avoided all mathematical symbols in his talk. He presented his work and part of that of his older friend and colleague from Norway, SOPHUS LIE, as the beginning of a program according to which geometry should be considered from the point of view of transformation groups which act on certain spaces. KLEIN's lecture joins a multitude of geometries, the theory of algebraic invariants, and even differential equations to group theory. It is a brilliant and much quoted work, and it was republished in a facsimile edition a century later. We cannot appraise here its actual influence, although it may be safe to say that the work of KLEIN's silent partner, SOPHUS LIE, became much more important than KLEIN's own work in the theory of transformation groups of spaces.

However, for the beginnings of combinatorial group theory, other aspects of KLEIN's work became important. The groups appearing in his lecture are not easily defined in terms of presentations and still today (1980), there does not exist a strong reason to do so. Even the concept of generators has to be modified in the sense of LIE's "infinitesimal substitutions" to be useful for the groups considered by KLEIN which, after all, do not even have a countable number of elements. It is true that countably infinite finitely generated groups had been considered at that time. They appear in a book on the theory of abelian functions by CLEBSCH and GORDON [1866] where, however, not even the *word* "group" is used, and later on in a series of papers on monodromy groups of homogeneous linear differential equations, the first of which seems to be a publication by C. JORDAN [1874].

All of these papers became significant for combinatorial group theory only much later. The main problem of this field is, after all, not the handling of the generators but that of the role of the defining relations. And this role became, literally speaking, visible, almost palpable, through the discovery of the symmetry groups of infinite tessellations of the non-euclidean plane.

Nowadays it is commonplace to say that group theory is the theory of symmetries. Indeed, to say that an object has a symmetry just means that it admits a transformation into itself, and the collection of all these symmetries is a group. Also, every group is a group of symmetries since it is a group of permutations of its own elements. But the word "symmetry" also has an everyday connotation which does not emerge if we consider the symmetries of finite sets, i.e., the symmetric groups of permutations. And finite geometric objects either have too few, namely, finitely many, symmetries as, for instance, a cube, or they have too many, namely a continuum of symmetries, like a circle or a sphere. But the tiling designs or tessellations of the euclidean or non-euclidean plane with a polygon as the basic tile have just the right structure to invite a description of their symmetry groups in terms of generators and defining relations. The generators and their inverses simply correspond to the edges of the basic polygonal tile, and the defining relations correspond to its vertices. If the polygon becomes more complicated, it may allow more and more interpretations as the fundamental region of a particular group of motions. A square can be interpreted both as the fundamental region of a group generated by reflections in its four sides or as the fundamental region of a group generated by two translations. But in both cases we immediately obtain a presentation of the group in terms of generators and of defining relations.

Reading the paper by DYCK [1882], there can be no doubt that it was this type of intuitive geometric insight which motivated his investigation. And DYCK, at this time, was KLEIN's assistant at the University of Leipsig, and KLEIN's work, induced by the problems of uniformization of algebraic curves through automorphic functions, had indeed been permeated for years with the theory of discontinuous groups of non-Euclidean motions; his important paper on RIEMANN's theory of complex variables quoted in Chapter I.3 was to appear one year after DYCK's paper.

The symmetry groups of tessellations helped to start combinatorial group theory but did not do much to develop it. One of the reasons for this is rather obvious: The simplicity of the original approach was deceptive. It seemed to be so much more natural to use geometric arguments than algebraic ones. Obviously, KLEIN saw no reason for developing an algebraic theory of free products, but the concept could have been abstracted from his 1883 paper. Another reason for the ineffectiveness of the geometric stimulus for the further development of combinatorial group theory may have been the question: Which infinite groups should one study? For the theory of finite groups, this problem never arose, and it is easy to see why. For the theory of infinite groups, one needed a motivation to start an investigation of groups which were significant in some sense and which could not be studied in a "concrete" manner as groups of transformations acting on a well-known mathematical object in a traditionally defined way. There can be no doubt that here POINCARÉ's discovery of the fundamental group of a space, together with the increasing interest in topology, was a decisive step. We do not know what led POINCARÉ to his discovery and, specifically, how much of the comtemporary theory of differential equations and of fundamental regions of groups which are discontinuous in their action on a three-dimensional space may have contributed to it. One cannot outguess a genius, and POINCARÉ himself had too many ideas to find the time for explaining how they arose. But although POINCARÉ provided the motivation for a development of combinatorial group theory, he contributed practically nothing to it. With due respect to TIETZE, whose work has been described in Chapter I.3, credit for this development must be given primarily to DEHN.

We wish to emphasize that the impact of POINCARÉ's discovery on group theory had several very important methodological consequences. First of all, CAYLEY's dictum, "a group is defined by means of the laws of combinations of its symbols," had now to be taken seriously. The object on which a group of transformations acts has disappeared completely if we define a group through a presentation. It is, of course, true that the theory of finite groups also had developed at this time (1910) in an abstract direction. In particular, many theorems could be and were stated in a purely abstract form. But neither the theory of permutation groups nor the theory of representations of groups as linear transformations of finite-dimensional vector spaces had been abolished by this development. On the contrary, the works of FROBENIUS, BURNSIDE, and L. E. DICKSON on group representations and on linear groups over finite fields must be considered as some of the most important contributions to the theory of finite groups which were made in the decades from 1870 to 1910. And the theory of Lie groups (or "continuous groups," as they were then called) had certainly not been separated from the concept of the groups as groups of transformations of a space. One may add here that POINCARÉ's definition of a fundamental group indeed starts with the action of a group on a space, but this fact is completely obliterated in the actual definition of the group in terms of generators and relations.

The second consequence of the emergence of combinatorial group theory was the fact that the fundamental problems of the new discipline could not be approached by employing the help of existing mathematical theories. First of all, analysis was no help whatsoever. It was discrete mathematics, namely, the problem of equivalence (or equality, as one used to say) of words of arbitrary but finite length in a finite or infinite alphabet under certain rules of change which were fixed in any particular problem but subject to a bewildering infinitude of choices. Of course, there existed one important branch of discrete mathematics, namely, number theory, and it played (and still plays) an important role in the abstract theory of finite groups. But in combinatorial group theory, it was, at best, of use only in special cases.

We should add here that topology itself had produced a new type of discrete, combinatorial mathematics. If one wishes, one can trace this back to EULER's solution of the Koenigsberg bridge problem. But it is certainly true and probably not incidental that DEHN himself (in his article on topology, written jointly with HEEGAARD) had done pioneering work in this direction. What appears to be incidental or, if one prefers, miraculous, is the fact that independent of DEHN and independent of topology, a contemporary mathematician had begun to ask questions of the type of the word problem in combinatorial group theory, but in an even more general and highly abstract setting. We are referring to the work of THUE, who may be considered as the founder of a general theory of semigroups. With one widely quoted exception, this work of his is largely forgotten nowadays. We do not know whether DEHN was influenced by THUE, and we have reasons to doubt it. We know that DEHN knew THUE personally, but only very superficially. DEHN mentioned THUE's work on occasion, observing that THUE's papers dealt with combinatorial problems. But he never used them, and indeed there is no known direct application of THUE's work to DEHN's group-theoretical problems.

The third effect of the emergence of combinatorial group theory was the increased emphasis on "decision problems" or "algorithms" or "general and effective methods." Again and again there appear in the work of DEHN the words: "to find a method of deciding in a finite number of steps" whether, e.g., a word in the generators of a group can be changed into the empty word by using the rules of change stated by the relations of a group. Certainly, this quest for "a method of deciding" is at least as old as Euclid's algorithm, and new emphasis had been given to it by HILBERT in the form of the tenth problem which he posed at the International Congress in Paris in 1900. But both EUCLID and HILBERT refer to problems of number theory, and this is a discipline with a tremendous wealth of far-reaching theorems. In combinatorial group theory, the quest for a decision procedure was *the* basic question, and not much help from known theorems was evident.

We know (through oral tradition) that DEHN was very much aware of the fact that the word problem in group theory had added a new aspect, a new type of questions, to mathematics. He was also aware of its difficulty. He used to say: "Solving the word problem for all groups may be as impossible as solving all mathematical problems." But there existed no possibility for DEHN in 1912 to anticipate the interaction of mathematical logic and group theory which occurred nearly half a century later and which was provoked by the word problem. (See Chapter II.11.)

We have already mentioned (in Chapter I.5) that DEHN's approach to combinatorial group theory was of a geometric rather than of an algebraic nature. Very successful algebraic methods were developed later, but only after the end of World War I. Although these methods were specific for combinatorial group theory, their emergence had been well prepared through the development of other parts of group theory and algebra.

Part of the progress in group theory during the period under consideration is due to a process which may be described as the streamlining of methods and concepts. This includes an increasing availability of abstraction. A good example for the effects of this process is given by a comparison of the paper by FROBENIUS and STICKELBERGER [1879] to Chapter 7 in the second edition of BURNSIDE's theory of finite groups which appeared in 1911. In both cases, the topic is the basis theorem for finite abelian groups. The paper by FROBENIUS covers 46 pages; the chapter by BURNSIDE is only 21 pages and includes part of the theory of automorphism groups of abelian groups of prime power order. FROBENIUS uses the theory of bilinear forms with integral coefficients and has a section on the residues of powers of integers with respect to a composite module. BURNSIDE uses an induction process with respect to quotient groups which had been introduced a few years earlier by HILTON. Both authors write with great clarity and do not omit details, but BURNSIDE had at his disposal previous chapters of his book which include the first Sylow theorem and the theory of composition series of a group. The contents of neither one of these chapters were completely available to FROBENIUS and STICKELBERGER at the time their paper appeared.

A glance at the table of contents in BURNSIDE's book [2nd edition, 1911] shows a large number of topics which were developed in the theory of finite groups but which could be extended at least partially to the theory of infinite groups including groups defined by generators and defining relations. We mention some of them, roughly in the order in which they appeared later in combinatorial group theory.

(1)   The automorphism group of a group. Inner and outer automorphisms.

(2)   The actions of the elements of a group on the right cosets of a subgroup. This leads to a group-theoretical characterization of the representations of a group as a transitive permutation group and to the theory of induced representations developed by FROBENIUS.

(3)   The concepts of characteristic subgroup, commutator subgroup, derived series, and solvability. The last three concepts came from the theory of Lie groups.

(4)   The abstract characterizations of direct products of groups of prime power order as nilpotent groups (or groups with a terminating lower central series).

(5)   The concept of a variety of groups, in this case appearing only as the set of finitely generated groups in which a fixed power of every element is the identity.

(6)   The Schur multiplicator.

(7)   Of indirect influence later on was the concept of the decomposition of a group into a direct product of groups.

Missing in BURNSIDE's book is the concept of the Frattini subgroup of a group which was discovered in 1885 (and which was transferred to infinite groups in 1937 by B. H. NEUMANN).

In addition to specifically group-theoretic developments, there were developments in other parts of algebra which were, in part, motivated by group-theoretical research but which, in turn, also influenced group theory. We mention, as cue words, only the terms: associative algebras, including group rings; representation modules; Lie algebras.

There is another reason why the conditions for purely algebraic research in combinatorial group theory were, after 1919, much more favorable than in 1882. It is a reason which exemplifies a general law in the emergence of new disciplines. It can be formulated by saying that group theory had come into its own. Groups had proved to be important for algebraic equations, for analysis (specifically for differential equations), for geometry, crystallography, number theory, and topology. Now (1919) they became interesting just because they were there. The beginning of this development can already be seen in groups in whose presentation any generator appears at most twice. It is true that DEHN sought a geometric interpretation for these groups, but the problem is an algebraic one. However, DEHN was not an algebraist. He was a geometer through and through, and he approached every problem with the vision and methods of a geometer. But the first mathematician to treat combinatorial group theory with purely algebraic methods was, in part, a student of DEHN. JACOB NIELSON lost his thesis advisor through death before he had finished his thesis, and the last part of his dissertation written at the University of Kiel and published in 1913 was based on a problem posed by DEHN. It was, of course, of a geometric nature, namely, a fixed-point problem. But NIELSEN's thesis was followed later by several purely algebraic papers which were motivated by DEHN's ideas. The results of these and the unfolding of combinatorial group theory as a discipline with its own methods and problems belongs in Part II of our historical study.

Chapter I.8

Modes of Communication. Growth and Distribution of Research in Group Theory

In this chapter we discuss the growth, distribution, and modes of communication of mathematical research in the period from 1880 to the end of World War II. We do not refer exclusively to research in a particular branch of mathematics, although some of the examples we give are taken from the literature quoted in the present book. We cannot confine ourselves to combinatorial group theory—there are comparatively few papers on this subject, and they have been mentioned and analyzed in the previous chapters. But we shall comment briefly on the spread of research in the theory of finite groups and we shall note that it differs from that in other disciplines. The modes of communication are well documented in the publications of the period under consideration (1880–1918), both in journals and books as well as in personal documents such as the letters of MINKOWSKI [1973] to HILBERT.

The mathematical world was much smaller in the period under consideration than in the following three decades. This refers not only to the number of mathematicians but also to their geographical distribution. Of the non-European countries with a great mathematical tradition in past centuries, only India had begun to contribute to the modern development which in Europe had started with the discovery of the calculus. And the Indian contribution was represented most visibly by the brilliant genius of SRINIVASA RAMANUJAN (1887–1920), whose work lies outside the scope of our investigation.

Mathematical research was mainly pursued in three regions, each of which was to some extent homogeneous with respect to the opportunities it offered its scholars. These are continental Europe, Great Britain, and North America (where the United States held nearly a monopoly at that time). Research from 1880 to 1918 was almost exclusively concentrated at universities. In the United States, this had been the case practically from the beginning. In England, the situation was more complicated. One is surprised to hear that ARTHUR CAYLEY (1821–1895) was a lawyer until 1863 when he became Sadlerian Professor of Pure Mathematics at Cambridge. But whatever developments may have taken place in Great Britain in the Eighteenth and early Nineteenth Century, they appear to have been of a gradual nature, and during the period from 1880 to 1918, the names of practically all of the prominent British mathematicians were associated with universities. The same was certainly true in continental Europe, but here a drastic change had taken place earlier under the influence of the French Revolution and the Napoleonic Era. In the Eighteenth Century, the academies had played an important role in mathematical research. LEONHARD EULER (1707–1783) was never associated with a university and JOSEPH LOUIS LAGRANGE (1736–1813) taught only at an artillery school from 1755 to 1766 in Turin, Italy—probably on a level well below that of his research—and became a member of a university (the Ecole Normale Supérieur in Paris) as late as 1795, at the age of 59.

We mention these facts because the universities opened lines of communication which are closed to the private scholar and available only to a limited degree to the academician through the contact with colleagues. Through contact between professor and doctoral student, the university offers an exceedingly effective channel for the propagation of ideas and the creation of schools. These effects are clearly visible in the continental European universities and in the graduate schools of the United States. Again, the situation was more complex in Great Britain, especially in England. There the doctorate was not a prerequisite for an academic career and some of the leading mathematicians of the country never acquired a Ph.D. degree, a circumstance which did not change until much later in the Twentieth Century. It is interesting to note that W. BURNSIDE was a Doctor of Science from Dublin and a Doctor of Laws from Edinburgh and was both an honorary Fellow of Pembroke College in Cambridge and a Fellow of the Royal Society but that he taught as a professor of Mathematics at the Royal Naval College in Greenwich. It is doubtful that he ever taught a course in group theory, and we are not aware of any Ph.D. students he might have had. He was, of course, very influential in other ways, partly through his papers and also through his book which is widely quoted and the reading of which induced P. HALL to work on group-theoretical problems. But the contrast between BURNSIDE and HALL is quite striking and illustrates a change in the role of the universities in England. HALL became a professor at Cambridge, and a large number of prominent English algebraists were his Ph.D. students.

Opportunities for personal contact outside of the universities were offered through the meetings of mathematical societies, invited colloquium talks, private visits, and international congresses.

According to G. A. MILLER [1935], by 1890 the following mathematical societies had been founded: The London Mathematical Society (1865), The Société Mathématique de France (1872), the Edinburgh Mathematical Society (1883), the Circulo Matematico di Palermo (1884), the American (originally, New York) Mathematical Society (1888), and the Deutsche Mathematiker Vereinigung (1890). MILLER [1935] does not mention the Moscow Mathematical Society which had been founded already in 1864. International congresses were held in the years 1897, 1900, 1904, 1908, and 1912, respectively, in Zuerich, Paris, Heidelberg, Rome, and Cambridge (England). We know of at least one instance where personal contact provided by such a congress led to a fruitful mathematical collaboration. POUL HEEGAARD from Copenhagen and MAX DEHN from Muenster (Germany) met at the congress in Heidelberg and started plans for a joint article on analysis situs (topology) for the German Encyclopedia of Mathematical Sciences which was published in 1907.

We do not have any quantitative data which would permit us to estimate the amount of travel undertaken for colloquium talks or private visits. One gets a general impression through the letters from MINKOWSKI to HILBERT, which were published in 1973. Certainly, travel was more expensive in relation to salaries at that time than it is today, and over long distances it was much more time consuming. The latter fact did not matter too much in Europe, and it may be noted that there existed no currency restrictions (the gold coins of France, Italy, and Switzerland were accepted in all three countries) and only Russia and Turkey required passports for visitors. However, crossing the Atlantic was a major enterprise. It was undertaken mainly in one direction: American students and scholars visited European universities for graduate or postgraduate studies. We know this from the biographies of American mathematicians and from the list of Ph.D. students in Goettingen. We are not aware of any instances where Europeans went to the United States to study. However, there were several cases where European scholars came to American universities as visiting professors or with permanent appointments. Of the people mentioned in the present book, FELIX KLEIN and H. MASCHKE are, respectively, examples of the first and second of these possibilities. That such visits or transfers were not more common was probably not due to a lack of American hospitality. The story of the negotiations between FELIX KLEIN and Johns Hopkins University (as told by REID [1978]) shows how difficult it was for Europeans to accept the different living conditions in America. It may be that a subconscious feeling that this was a strange country also played a role. It is somewhat curious to read in the travel guide for the United States, published by Baedecker (Leipzig) in 1909 not only a warning that America was expensive by European standards but also an assurance that, even in the western parts of the country, it was completely out of place to carry a handgun. (Of course, such an assurance was unnecessary for people like KLEIN who knew Americans.)

Within Europe, transfer of mathematicians across national boundaries was at least comparatively frequent, although not really common. In the first place, Switzerland and, to a lesser but still noticeable degree, Germany, occasionally filled university positions with foreign scholars. In Germany, even citizenship went automatically with a tenured position. Of the mathematicians mentioned in the present paper, MINKOWSKI was a professor at the Eidgenoessische Technische Hochschule in Zuerich from 1896 to 1902 and HURWITZ went there permanently in 1892. The Norwegian SOPHUS LIE held a professorship in Leipzig from 1886 to 1898 and the Russian-born ISSAI SCHUR was appointed as a professor at the University of Bonn in 1911.

Whatever the effect of personal contacts through common places of work, travel, and international exchanges may have been, we have very little documentary evidence for it. The documentation of the fact that the encounter between DEHN and HEEGAARD at Heidelberg in 1904 was a very fruitful one is an exception and it had been transmitted through an oral communication from Mrs. TONI DEHN. However, it seems reasonable to assume that personal contacts were important in many other cases as well.

Undoubtedly the most effective and the most common mode of communication was that provided through publications. For these, there existed, from 1826 on, a new and highly efficient vehicle: mathematical journals. In the Eighteenth Century, mathematicians had to rely on books and on the publications of academies. These continued to play a role. But the increasing volume of research and the increasing specialization of scholars made the publications of the academies a rather cumbersome instrument of communication. There might be not more than three or four mathematical papers in a heavy volume published by an academy. As a rule, a university would not subscribe to more that one copy of these works which then had to be used jointly by mathematicians, linguists, and historians. (Even the separation of papers into two classes, catering respectively to the sciences and the humanities took place mostly after 1900 or even after 1918.)

The first two mathematical journals were the *Journal fuer die reine und angewandte Mathematik* (for nearly a century called *Crelle's Journal* after its founder A. L. CRELLE) and the *Journal des mathématiques pures et appliqués* (also called *Liouville's Journal* after its founder, J. LIOUVILLE). Both of them had private (i.e., commercial) publishers, although until 1918 Crelle's Journal mentions on its title page the support of high Royal Prussian Authorities (*Die thaetige Befoerderung hoher Koeniglich Preussischer Behoerden*). This support consisted, at least originally, in subscriptions from the military schools in Prussia. Crelle's Journal was founded in 1826 and Liouville's Journal first appeared in 1838. Later, mathematical societies and even individual universities began to publish mathematical journals. In fact, commercially published journals existed, at least until 1914, only in Europe. The number of journals which published mathematical papers increased steadily. The *Jahrbuch ueber die Fortschritte der Mathematik* lists in its first volume (which appeared in 1872) about 80 sources of mathematical publications and has 426 pages. The volume covering 1913 lists about 160 such sources, including an Indian and a Japanese journal, and had 1210 pages. (Of course, not all mathematical publications appeared in mathematical journals).

Language barriers do not seem to have played any role, at least not for the research mathematician. The use of Latin as the language of scholars had died out in the middle of the Nineteenth Century, even in Germany, where C. F. GAUSS and K. G. J. JACOBI were probably the last of the prominent mathematicians to use it. But it was still a universally known language among scholars, and the older literature was therefore still accessible. However, the overwhelming majority of mathematical papers used one of the four languages admitted at international congresses, namely, English, French, German, and Italian. Judging from the abundance of references in the literature to papers not written in the language of the author but in one of these four languages, they seem to have been understood everywhere at least when used in mathematical publications. This remark refers to active scholars. For students, the situation may have been different. In his monograph, BURNSIDE [1911, p. 506] notes with regret that there existed at that time no English textbook on algebraic number theory. He mentions HILBERT's *Zahlbericht* (HILBERT, [1897]) as a source and adds that this work had been translated into French. Of the four languages mentioned above, this was certainly the one most widely known. As an illustration, we may note the fact that the numerous passages of conversations in French which appear in some of TOLSTOY's novels were left untranslated in German editions.

It was easier to publish long papers than in later periods. Expository sections appeared with much higher frequency in the publications of that time then now, and the same is true for elaborate calculations and the statement of explicit formulas. When MITTAG-LEFFLER founded *Acta Mathematica* in 1882, he stipulated that "no paper shall ever be rejected merely because of its length."

Apart from the journals themselves, the offprints of published articles given to their authors were probably considerably more important to the dissemination of results than they are today (1980). At least one firm, the Buchhandlung Gustav Fock in Leipzig, bought collections of offprints and sold them individually. MAGNUS still owns offprints of papers by FRICKE, FROBENIUS, and SCHUR which had appeared in the less accessible journals like the *Sitzungsberichte der Preussischen Akademie* or the *Nachrichten der*

Gesellschaft der Wissenschaften zu Goettingen which he bought there before 1933. The importance of offprints was due to the lack of easily available and inexpensive reproduction methods. We do not know the exact year when photostatic copies became available, but this certainly happened well before the year 1900. However, they were expensive. [One preferred to order the inconvenient negative (white print on black background) because it was only about half as expensive as the positive.] Also, one had to wait at least a day for delivery after ordering the copy. Preprints were practically unknown, although the mimeographing method, too, existed before 1900. Even the typewriter, which could produce at least two copies, was used very little by mathematicians because formulas had to be inserted by hand. Until World War II, most mathematical journals accepted handwritten papers for publication.

Letters were, of course, a common means of communication, although long letters with the exposition of elaborate proofs were certainly rare. Combined with the use of journals, they produced a very effective means of communication. An example involving authors of three nations is the following one: DYCK [1883] acknowledges a letter from Mr. WASSILIEFF in Kasan (Russia), informing him that his presentation of the group of the icosahedron, as published in DYCK [1882] had already been found by HAMILTON in 1856. HAMILTON's terminology (he speaks of "roots of unity" rather than of "group elements") is not the one used by CAYLEY and DYCK, but the result is indeed the same.

Systematic presentations of the methods and results of a particular discipline certainly existed for a long time, even before the appearance of EUCLID's *Elements*. In group theory, the first systematic exposition of this field is the *Traité des substitutions* by CAMILLE JORDAN which appeared in 1870. It was followed by numerous textbooks which either contained chapters on group theory like the *Lehrbuch der Algebra* by H. WEBER [1896] or were dedicated entirely to group theory like the *Theory of Groups of Finite Order* by BURNSIDE [1897a]. These works played an important role and they were used not only as textbooks but also as works of reference. But the increasing volume of mathematical knowledge stimulated a novel effort to organize information and to make it more easily and efficiently available. This was the purpose of reviewing journals and of surveys on various levels of sophistication.

Volume 1 of the *Jahrbuch ueber die Fortschritte der Mathematik* reports on the papers published in 1868. It was the first reviewing journal in mathematics and, until 1930, the only one. It was published by GEORG REIMER in Berlin. All the reviews were in German, in spite of the fact that German journals published papers in French, Italian, and English as well. Its importance as a source of information was (and still is) considerable. The minimum amount of information given about a publication is the full title, the exact reference (including page numbers) and, in the case of books, the place and language of publication. The editors even made a serious effort to compensate for the lack of information during World War I. Volume 48 (1920) contains a special section reviewing Russian papers which had appeared during the war years. The reviews were arranged according to fields. Group theory had a special section after 1897. Because the reviews of all the publications of a full year are listed in one section, it is comparatively easy to browse through the literature of a decade in a particular field. However, there are no cross references, and one may miss papers containing group-theoretical results which are listed under other headings—e.g., MINKOWSKI [1905].

The quality of the reviews varies widely. In group theory, at least three quarters of the reviews are fair and give an idea of the contents and importance of the paper. Also, many reviews refer to earlier relevant papers. Since the total number of reviewers was, at least until 1918, rather small (not more than 60 for all of mathematics), one observes that the quality of the review depends mainly on the particular reviewer. According to the old maxim *De mortuis nil nisi bene* (Say nothing about the dead unless it is something good), we shall not mention the incompetent, the lazy, or the snobbish reviewers but one of the writers of the present book (MAGNUS) feels that this is an opportunity to record an acknowledgment of the excellent reviews written by A. LOEWY.

The need for surveys of the literature, including surveys of results, was met mainly by the *Berichte* (reports) published by the *Deutsche Mathematiker Vereinigung* (German Mathematical Society) and the *Enzyklopaedie der mathematishen Wissenschaften* (*Encyclopedia of the Mathematical Sciences*). The most famous of the German Mathematical Society reports dealing with the theory of algebraic number fields and written by HILBERT in 1897 is actually a monograph containing complete proofs of most of the theorems and also new results. But nothing like this exists for group theory. The report by SCHLESINGER [1909] contains a very large number of references to papers dealing with group-theoretical investigations arising in the theory of ordinary differential equations, classifying these papers by the symbol [s] (for "substitution"). But the text reveals little information which would be important for a group theorist. An American report on the literature which is dedicated entirely to the theory of groups is due to EASTON [1902] and was published in Boston. It is a bibliography with information about results in a very condensed form.

The *Encyclopedia of the Mathematical Sciences* contains two articles on finite groups, both in Volume 1, ca. 1900. They were written by H. BURKHARDT (*Endliche discrete Gruppen*) and by A. WIMAN (*Endliche Gruppen linearer Substitutionen*). Volume 2 contains an article (No. IIB. 4) by R. FRICKE on automorphic functions, including elliptic modular functions, which contains a good survey of this field including references to some group-theoretical results. FRICKE's article appeared in 1913.

The rapid development of the mathematical sciences during the period under consideration made some articles in the *Encyclopedia* obsolete after a comparatively short time. For this reason, a French encyclopedia entitled *Encyclopédie des sciences mathématiques pures et appliqués* was started in about 1910. It was never completed, due to the interruption caused by World War I. The article on group theory breaks off in the middle of a paragraph. Although the published part dealt only with rather basic results of group theory, it was longer than the articles by BURKHARDT and WIMAN taken together. Today, the incomplete work seems to be nearly inaccessible. It is not even mentioned in the list of publications in the *Fortschritte der Mathematik*. However, its existence can be guaranteed by one of the authors of this essay (MAGNUS), who once owned the published portion.

By far the best survey article on group theory written before World War I appeared in 1910 in the second edition of a *Repertorium* which had been started by E. PASCAL who was a professor at the University of Naples in Italy. The article on group theory in the second edition is written by A. LOEWY. In spite of its condensed form it is very clear, well organized, up to date, and prefaced by a careful exposition of the historical development with many references.

References in the literature before 1914 (and even before World War II) were, to a large extent, in a deplorable state. This was not only due to the fact that many authors did not mention relevant literature. This still happens. But the form of the references is, by today's standards, inadequate in the majority of papers, even in the *Encyclopedia*. Neither the title of the paper nor its length appear in most references. Even the year of publication is frequently missing. Many references give only the abbreviated name of the journal and the number of the first page of the paper. The situation becomes particularly irritating for the reader when a large number of references is given. In the *Encyclopedia* all references are in footnotes. Since every reference appears only once, the footnotes contain references to other footnotes. The name of the author is frequently not given in the reference but only in the text. This style for quoting the literature was mandatory for the authors of the articles of the *Encyclopedia*, even in the second edition (which began to appear after 1930 and was never completed). But in publications in journals and monographs, the authors could do as they pleased. And here today's readers will appreciate authors like BURNSIDE, FRICKE, LOEWY, G. A. MILLER, and others who stand out among their contemporaries by giving detailed references.

We conclude this section with a few tentative statements about the propagation of research and ideas which are based on a nonsystematic collection of observations.

The interest in group theory, at least as far as finite groups are concerned, seems to have spread rather quickly in the period from 1870 to 1914. The list of textbooks in the bibliography in WUSSING [1969] shows the international character of this interest. The same is not true for other fields. Specifically, the theory of algebraic number fields was pursued mainly on the European continent during the same period, in spite of the fact that it goes back to ERNST EDUARD KUMMER (1810–1893) and received a powerful impetus through the *Zahlbericht* of DAVID HILBERT in 1897. Its applications to group theory were well known, both in England and in America, but we have already mentioned BURNSIDE's remark that no textbook in English existed for this field as late as 1911. Other fields, e.g., algebraic geometry, seem to have had a similar fate.

We are not aware of a lack of recognition of the work of important authors in group theory, although it may be said that the work of T. MOLIEN (1861–1941) on the theory of representations of finite groups had been overshadowed unfairly by that of BURNSIDE, FROBENIUS, and SCHUR. Of course, disregard of authors has occurred in other fields. As an example, we mentioned A. M. LIAPUNOV (1857–1918). That his important paper on the stability of motion was not recognized when it appeared in Russian is understandable. Very few people outside of Russia could have read it at that time. But it appeared again in French under the title *Problème général de la stabilité de mouvement* in the *Annales de la Faculté des Sciences de l'Université Toulouse*, Ser. 2, Vol. 9, pp. 203–474 in 1907 without being recognized by people working in the field. The final international recognition came only after World War II when the Princeton University Press republished it in 1947 (jointly with Oxford University Press).

The influence of leading mathematicians on their students is, for the period under consideration, very hard to appraise. Undoubtedly, FELIX KLEIN had a very strong influence on the work of his students VON DYCK and FRICKE. And the influence of HILBERT on his students is well known, although several of them, including DEHN, later worked in fields different from that of their Ph.D. thesis. But TIETZE was not a student of POINCARÉ, although his long paper of 1908 is completely motivated by POINCARÉ's work. (TIETZE received his Ph.D. in Vienna in 1904 with a Ph.D. thesis on functional equations the solutions of which cannot satisfy any algebraic differential equation.)

It is probably safe to say that, at this time, it was easier to begin work in a new field than it is today. This statement is confirmed by the fact that quite a few of the prominent mathematicians mentioned in our study had done important work in several fields. FROBENIUS is mainly remembered as a algebraist, but he started as an analyst and nearly a third of this work is dedicated to analysis. There are quite a few other examples of this type, and there is even the astonishing case of L. E. DICKSON whose work is mainly in group theory but who nevertheless found the time to write his monumental *History of the Theory of Numbers* which appeared in three volumes totaling more than 1600 pages. (It is still in print, having been republished in 1971 by the Chelsea Publishing Company, New York.) On the other hand, we already have the phenomenon of the "chemically pure" specialist, represented in group theory by G. A. MILLER (1863–1951) whose 359 papers deal almost exclusively with finite groups

As a last observation, we mention the 1901 Ph.D. thesis of W. BOY (described in BOY [1903]). It contains the proof of a result in topology which is now known as the Whitney–Graustein theorem. The surprising thing here is that BOY's thesis advisor was DAVID HILBERT. HILBERT contributed to practically all fields of mathematics, but topology is the most prominent exception. It is certainly very unlikely today that a Ph.D. advisor would assign a topic in a field in which he or she is an outsider. Incidentally, we are not aware of any similar case even from that time.

We know that the observations made in the last part of this chapter give, at best, a rather sketchy picture of the situation. We hope to be able to put it into sharper focus in our report about a later period.

Chapter I.9

Biographical Notes

This chapter contains references to obituaries of mathematicians mentioned in the present book whose papers appeared before 1918. If the obituary quoted contains a bibliography, this is mentioned. In most cases, it is not included in the obituary quoted. In several cases, particularly those of Russian authors, there may also exist obituaries with a bibliography which we have been unable to locate.

There are three sources of obituaries (usually without complete bibliography) which appear repeatedly. They are:

World's Who's Who in Science, First Edition, 1968, published by Marquis–Who's Who, 200 East Ohio Street, Chicago, IL 60611. U.S.A.

Publications of the *Leo Baeck Institute*, New York, Yearbook 18, 1973; published for the Institute by Secker and Warburg, London. This issue contains obituaries of German mathematicians of Jewish descent who were victims of Nazi persecution, 1933–1945.

J. C. POGGENDORFF's *Biographisch-Literarisches Handwörterbuch*, Vol. II–IV, published by Johann Ambrosius Barth, Leipzig, Vol. V–VI, Verlag Chemie, Berlin, Vol. VIIa, Akademie Verlag, Berlin.

These sources will be referred to, respectively, by the letters (W.W.), (L.B.I.) and (P.).

BAKER, HENRY FREDERICK, 1866–1956 (W.W.).
BIANCHI, LUIGI, 1856–1928 (W.W.).
BIEBERBACH, LUDWIG, 1886–? (W.W.).
BLUMENTHAL, OTTO, 1876–1944 (L.B.I., p. 157).
BOY, W. We know nothing about him, not even his first name, except for the fact that he was a
    Ph.D. student of DAVID HILBERT.
BURNSIDE, WILLIAM, 1852–1927 (W.W.).
CAMPBELL, J. E., 1862–1924 (P.).
CANTOR, GEORG FERDINAND LUDWIG PHILIPP, 1845–1918 (W.W.).
CAYLEY, ARTHUR, 1821–1895 (W.W.). *Collected papers of Arthur Cayley*, 12 volumes, Cambridge University Press, Cambridge, England, 1896.

CLEBSCH, RUDOLF FRIEDRICH ALFRED (A. Clebsch in publications), 1833–1872 (W.W.).

DEHN, MAX, 1878–1951. Obituary with bibliography: MAGNUS and MOUFANG [1954]. Detailed biography in SIEGEL [1965]. Memorial essay on occasion of his hundredth birthday: MAGNUS [1978].

DE SÉGUIER, J.-A., 1862–1937. We do not even know his first name. The information we have (that he was a private scholar) and the dates are taken from WUSSING [1969].

DICKSON, LEONARD EUGENE, 1874–1954 (W.W.).

DYCK, WALTHER VON, 1856–1934 (W.W.).

FRATTINI, G, 1852–1925 (P.).

FRICKE, ROBERT, 1861–1930 (P.).

FROBENIUS, FERDINAND GEORG, 1849–1917 (W.W.). Collected papers: *Ferdinand Georg Frobenius, Gesammelte Abhandlungen* (ed. J.-P. Serre), Springer–Verlag, Berlin-Heidelberg-New York, 1968 (Three volumes).

GIESEKING, HUGO, Ph.D. thesis, 1912. Died in World War I.

GRAVÉ, DMITRI ALEKSANDROVICH, 1863–1939. Biography by A. P. Yushkevich, *History of Mathematics in Russia until 1917*, Nauka, Moscow, 1968.

HAMILTON, SIR WILLIAM ROWAN, 1805–1865 (W.W.). Collected papers: *The Mathematical papers of Sir William Rowan Hamilton* (3 volumes), Cambridge University Press Cambridge, England, 1930–1967; obituary in Vol. 1, pp. IX–XVI.

HAUSDORFF, FELIX, 1868–1942 (L.B.I., p. 165).

HEEGAARD, POUL, 1871–?.

HILBERT, DAVID, 1862–1943 (W.W.). Detailed biography: *Hilbert*, by Constance Reid, Springer-Verlag, New York-Heidelberg-Berlin, 1970, 290 pp. Collected papers: *David Hilbert*, Gesammelte Abhandlungen, 3 Volumes, Verlag Julius Springer, Berlin, 1932–1934.

HOYER, P. We do not know his first name or any dates. W. GASCHÜTZ has informed us that HOYER was a high school teacher.

HUMBERT, MARIE GEORGES, 1859–1921 (W.W.).

HURWITZ, ADOLF, 1859–1919. Biography and collected papers in: *Mathematische Werke von Adolf Hurwitz*, 2 volumes, Birkhäuser Verlag, Basel, 1932. Obituary by D. Hilbert: *Nachrichten der Gesellschaft der Wissenschaften zu Goettingen*, 1920, pp. 1–9.

JORDAN, CAMILLE, 1838–1922 (P.).

KLEIN, CHRISTIAN FELIX (Felix Klein or F. Klein in publications), 1849–1925 (W.W.). Bibliography: *Felix Klein, Gesammelte mathematische Abhandlungen*, 3 Volumes, Julius Springer, Berlin, 1921–1923.

LIE, MARIUS SOPHUS, (S. Lie or Sophus Lie in publications), 1842–1899 (W.W.). Bibliography: *Sophus Lie, Samlede Abhandlinger/Gesammelte Abhandlungen*, 6 Volumes, H. Ashehoug, Oslo/B. G. Teubner, Leipzig, 1927–1934.

LISTING, JOHANN BENEDICT, 1808–1882 (P.).

LOEWY, ALFRED,1873–1935. (L.B.I., p. 170).

MASCHKE, H. 1853–1908 (P.).

MILLER, GEORGE ABRAM, 1863–1951 (W.W.). Collected papers: *The Collected Works of George Abram Miller*, 4 volumes, University of Illinois, Urbana, IL, 1935–1955.

MINKOWSKI, HERMANN, 1864–1909 (W.W.). Obituary by D. Hilbert, *Nachrichten der K. Gesellschaft der Wissenschaften zu Göttingen*, 1901, 1–30. Collected papers: *Gesammelte Abhandlungen von Hermann Minkowski*, 2 volumes, Leipzig, 1911; reprinted Chelsea Publishing Co., New York, 1967.

NIELSEN, JACOB, 1890–1959. Obituary with bibliography: Werner Fenchel, *Jacob Nielsen in Memoriam*, Acta Math. **103**, Issue 3–4, pp. VII–XIX, 1960.

NOETHER, EMMY, 1882–1935 (W.W.). Biography and bibliography by Auguste Dick, *Emmy Noether*, Birkhäuser Verlag, Basel, 1970, 72 pp.

PICARD, CHARLES EMILE (E. Picard in publications), 1856–1941 (W.W.).

PICK, G., 1859–1943(?) (P.).

POINCARÉ, JULES HENRI (H. Poincaré or Henri Poincaré in publications), 1854–1912 (W.W). Volume 38, (1921) of *Acta Mathematica* is dedicated to his memory. Collected papers: *Oeuvres de Henri Poincaré*, 11 volumes, Gauthier Villars, Paris, 1928–1956.

REMAK, ROBERT ERICH, 1888–ca. 1942 (L.B.I., p. 176).

RIEMANN, BERNHARD, 1826–1866 (W.W.). For biography and collected papers see RIEMANN [1892] in list of references.

SCHLESINGER, LUDWIG, 1864–1933 (P.).

SCHMIDT, OTTO (also, O. J. Schmidt in publications), 1891–1956 (W.W.).

SCHUR, ISSAI, 1875–1941. Biography and collected papers in: *Issai Schur, Gesammelte Abhandlungen*, 3 volumes, Springer-Verlag, New York-Heidelberg-Berlin, 1973.

THUE, AXEL, 1863–1922. Obituary and partial bibliography in: *Selected Mathematical Papers of Axel Thue*, Universitets Forlaget, Oslo-Bergen-Tromsø, 1977.

TIETZE, HEINRICH, 1880–1964. Obituary by G. Aumann in *Bayerische Akademie der Wissenschaften*, Jahrbuch, 1964, pp. 197–201.

VESSIOT, E, 1864–1933 (P.).

VOGT, H. ?–?.

WEBER, HEINRICH, 1842–1913 (W.W.).

WIRTINGER, W., 1865–1945.

Chapter I.10

Notes on Terminology and Definitions

It is part of the purpose of every historical essay to make it easier to go back to the sources for those readers who may be interested in doing so. Since we are dealing with the history of rather recent times, this ought to be an easy task and we believe that the remarks made below will be all that is needed. However, the fact that we are dealing with recent times also introduces a peculiar difficulty. The number of specialized technical terms which we need increases rapidly as we approach the present and the question arises how many of them we will have to define. The problem is aggravated by the fact that we need concepts from fields other than group theory as well. In the second part of this chapter we give some definitions and explain why we replaced others by references to the literature. Finally, we explain a few notations used in the text.

In group theory, as in other fields, the terminology has changed since the time at which our historical account begins. Except for the few cases where we quote the authors verbatim, we have always used current (1980) terminology. Fortunately, the difficulties arising for today's readers of older papers from a change of terminology are not very great since the habit of defining the terms used in a paper was much more widespread in the past than it is today. Even the linguistic difficulties for readers who need a dictionary for languages different from their own are only moderate ones in those cases where the terms used are derived from Latin or Greek roots. They are slightly increased for readers of German papers because of a spelling reform carried out in Germany in 1900 which in many cases replaced the letter "c" by the letter "k" and dropped the "h" in many words with a "th." In addition, several words which are practically identical in English, French, and Italian, because of a common Latin root, are very different in German. We mention here only two that appeared frequently in group theory:

> divisor, diviseur, divisore, Teiler
> distinguished, distingué, distinto, ausgezeichnet.

The only terms used frequently in the older literature which we believe require identification or comments are the substitutes for the words *subgroup* and *normal subgroup* and the meaning of the words *isomorphism*, *discrete*, and *discontinuous*. A subgroup is frequently called a divisor, and a normal subgroup is also called a distinguished or invariant subgroup or divisor. The term "isomorphism" includes in the older literature what is now called a homomorphism. Where a distinction becomes necessary, the epithets "holohedral" and "merohedral" may be used. Also, "isomorphism" is sometimes (e.g., in BURNSIDE [1911]) used for what is now called an automorphism. The term "homomorphism" appears already in DE SÉGUIER [1904] but not in BURNSIDE [1911]. It had been introduced by F. KLEIN. Probably the terms most confusing for a modern reader are those of "discrete" and "discontinuous." Up to about 1940, they were used mainly for infinite (mostly countable) groups which are not Lie groups (which then were called "continuous groups"). However, this usage is not a uniform one. Whenever groups of selfmappings of a topological (in particular, a metric) space are considered, the term "discontinuous" is nearly equivalent to the term "discrete" in modern usage, and "properly discontinuous" was used for what is now called discontinuous. These remarks at least apply to the definitions given on pp. 61–63 in the standard work of FRICKE and KLEIN [1897]. In our text, we use the words "discontinuous" and "discrete" (except for quotations and titles of papers) exclusively in the following sense: A group G acting as a group of invertible selfmappings of a topological space S is called discontinuous at a point P of S if there exists a neighborhood N of P such that only finitely many elements of G will map P onto a point of N. And the term "discrete" is reserved for a group G of matrices with entries from the field of complex numbers such that there exists no infinite sequence of distinct matrices in G for which the sequences of all the entries converge. Incidentally, a sequence of this excluded type is called an "infinitesimal substitution" by FRICKE, KLEIN, and others.

The term *fundamental region* of a discontinuous group G always denotes an open set with some of its boundary such that under the action of G every point in S is the image of at least one point of the fundamental region, but G cannot map any point of the open set into another one. The images of the fundamental region under the action of the group G then fill the space S (or that part of S where G is discontinuous) without gaps and overlappings.

The concept of a fundamental region is the most important nonelementary non-group-theoretical concept appearing in Chapters I.1–I.5 and I.7. We do not give a definition for the nonelementary concepts appearing in Chapter I.6 which is a rather sketchy survey and as such makes sense only for readers who are somewhat acquainted with the topics mentioned there. However, some of the group-theoretical concepts appearing in Chapter I.6, e.g., that of the lower central series, will be defined in Part II which will discuss the history of later periods. Right now we shall give only a concise set of definitions of terms appearing in Chapters I.1–I.5 and I.8 which are specific for combinatorial group theory. Terms from other fields appearing in these Chapters are assumed to be known or else are sufficiently complex to be unsuitable for a listing in this Chapter. Some of them are covered by references.

A *presentation* of a group G is defined as follows: We are given a set (not necessarily finite or even countable) of pairs of symbols a_ν, a_ν^{-1}. A finite sequence of these symbols is called a *word* and the number of symbols in a word is called its *length*. The empty word (with no symbols) is denoted by 1. If W_1, W_2 are words, their *product* $W_1 W_2$ is defined by juxtaposition. The symbols a_ν and a_ν^{-1} are called *inverses* of each other and the *inverse* W^{-1} of a word W is obtained by replacing each symbol in W by its inverse and then reversing the order of the symbols. The words $a_\nu a_\nu^{-1}$ and $a_\nu^{-1} a_\nu$ are called *trivial relators*. Apart from these, we can select an arbitrary set of words R_λ which are called *defining relators*. Two words are called *equivalent* if they can be changed into each other by insertion or deletion of subwords which are relators or their inverses. The equivalence classes of words form a group G with the equivalence class of 1 as the unit element or identity. Any particular word W *defines* an element g of G, namely, the equivalence class to which W belongs. With a slight relaxation of the requirements of precision, we call the a_ν (and not their equivalence classes) the *generators* of G and we write $R_\lambda = 1$ and call this the *defining relations* of G. The combination of a_ν and R_λ is called a *presentation* of G. The terms "finitely generated," "finitely related," and "finitely presented" are self-explanatory. A group G is called *free* if it has a presentation with generators a_ν and an empty set of (nontrivial) defining relators. The a_ν are then called *free generators*. A word is called *freely reduced* if it contains no subwords which are trivial relators. It is called *cyclically reduced* if, in addition, its first and last symbols are not inverses of each other. The number of free generators is called the *rank* of a free group.

The *free product* of two groups A and B with presentations a_ν, R_λ and b_μ, S_λ can be defined as the group G with generators a_ν, b_μ and defining relators R_ρ, S_σ. (Here the R_ρ are words in the a_ν and the S_σ are words in the b_μ.) We denote G by $A * B$ and observe that (trivially) every element $g \neq 1$ of G can be represented by a word of one of the forms

$$\alpha, \beta, \alpha\beta, \alpha_1\beta_1\alpha_2\beta_2 \cdots \alpha_K\beta_K \qquad (K \geq 2),$$

where $\alpha, \alpha_1, \ldots, \alpha_K$ are words in the a_ν, $\beta, \beta_1, \ldots, \beta_K$ are words in the b_μ and where $\alpha, \beta, \beta_1, \alpha_2, \beta_2, \ldots, \alpha_K$ are words which do not define the unit elements in A or in B. The first theorem about free products (sometimes called the *existence theorem* for free products) then states that none of the words listed above equals the unit element in G. These words are then called the *normal form* of the elements they represent. A free group is then the free product of the infinite cyclic groups generated by its distinct generators and its elements $\neq 1$ have the normal form of Proposition 1 of Chapter I.2.

The term *torsion free* denotes a group in which the identity is the only element of finite order.

We conclude this chapter with a few remarks about the notations we use. The symbols $\mathbb{Z}$, $\mathbb{R}$, $\mathbb{C}$ denote, respectively, the integers, real numbers, and complex numbers. The symbols $GL(n, R)$, $SL(n, R)$, and $PSL(n, R)$ denote, respectively, the group of nonsingular $n \times n$ matrices (general linear group), the group of $n \times n$ matrices with determinant $+1$ (special linear group), and the projective special linear group over a (commutative) integral domain R with unit element. The symplectic linear group of $2n \times 2n$ matrices with entries from R is denoted by $Sp(2n, R)$.

… <p>

Chapter I.11

Sources

Quite a few of the papers published before 1918 and listed in our Bibliography are also quoted in the work by H. WUSSING [1969] entitled *Genesis des abstrakten Gruppenbegriffes* (*Genesis of the Concept of an Abstract Group*). Although this looks like a topic of rather limited scope, WUSSING's book has 258 pages and gives 747 references. It lists a very large number of occasions where, in a variety of mathematical disciplines, the group concept turned out to be of importance, and it analyzes their effect on the development of that concept. As far as the topic of WUSSING's book is concerned, it is the most nearly perfect and complete account of the facts in a chapter of the history of mathematics one may hope to obtain.

We cannot claim the same degree of completeness for our account. One of the reasons for this has already been mentioned in Chapter I.1. The ways in which ideas and results were transmitted have usually not been documented for the period under consideration. But even our report on the facts, i.e., on the published results, may have gaps. The reason for this is mainly the overwhelming amount of material to be sifted. Several thousand papers on group theory were published during the period from 1882 to 1918, and we have already indicated in Chapter I.8 that even the papers listed in the chapters on group theory in the *Fortschritte der Mathematik* do not contain all of the relevant information. And since we certainly could not and did not read all mathematical papers, we shall at least try to describe briefly what we actually did do.

Up to a point, we followed the advice given by EMMY NOETHER to MAGNUS in 1934 when he accepted the task of writing an article on "general group theory", i.e., group theory with the exclusion of the theory of Lie groups and of group representations and linear groups for the second edition of the German encyclopedia of mathematical sciences (MAGNUS [1939]). She said: "First write down what you know. Then check the literature and expand." It is the checking of the literature which has to be described. At that time, MAGNUS read all the reviews in the *Fortschritte der Mathematik* on group theory up to 1936. After that, the question arose: What is important? The same question arose, of course, for the writing of the present book, but in our case the problem was even more difficult since we are concerned with the relations between papers, and here the old reviews offer even less help than for a project of recording the facts (i.e., the theorems). Therefore, we started with the obvious and well-known papers by DYCK, TIETZE, and DEHN, with the important monographs by FRICKE and KLEIN, BURNSIDE, and DE SÉGUIER, the encyclopedic surveys including the one by G. A. MILLER and, of course, the book by WUSSING. We also consulted the collected papers of FROBENIUS, I. SCHUR, MINKOWSKI, and of HURWITZ, who is a particularly conscientious author with respect to references. And we wish to acknowledge the valuable suggestions and pieces of information we received from colleagues with historical interests.

All of this does not guarantee completeness. But we hope to have discovered and recorded at least one example for every typical phenomenon of the transmission of ideas and referred at least to the dominant theorems of the field found before 1918.

Part II

The Emergence of Combinatorial Group Theory as an Independent Field

Chapter II.1

Introduction to Part II

This part of our book deals mainly with the developments during the period from 1918 to 1945, that is, from the end of World War I to the end of World War II. The reasons for our choices of these dates are twofold and are not based on the fact that both wars reduced the production of research in pure mathematics rather sharply. Starting with NIELSEN's papers which began to appear in 1918, problems of combinatorial group theory were investigated and solved which do not show an obvious dependence on problems in topology or in other fields. This does not mean that problems of this type had ceased to have a stimulating effect on combinatorial group theory. But then the field began to develop both its own problems and its own methods, and they were dealt with and developed by a new generation of authors. In the years after 1945, this phenomenon continued but in a much less pronounced way. The main change which occurred after this date was the acceleration in the rate of growth of the literature. At present, combinatorial group theory constitutes an important and pervasive ingredient in the theory of infinite groups with the exclusion of Lie groups and abelian groups. If we include the latter, the literature reviewed in the Mathematical Reviews during the period from 1940 to 1970 consists of 4563 publications. BAUMSLAG [1974] has classified the reviews of these in 24 chapters and 264 sections and has supplemented the classification with cross references in a two-volume work of more than 1000 pages. Obviously, no historical narrative can fully cover that much material, and we now proceed with a description of our method for selecting the topics for Part II of our book.

For the period from 1918 to 1937, we chose seven topics which we consider to be the most important contributions to our field. The arrangement of the corresponding seven chapters is chronological with respect to the first papers quoted in each section. But after that we refer to the later literature on the topic under discussion with the purpose of telling a

straightforward story. How far we pursue this story and how much literature we quote is a matter of expediency. We have tried to avoid the
formulation of highly technical results, and we refer, in all cases, to surveys
or monographs covering the later literature. Although the number of papers
published during the two decades from 1918 to 1937 is still manageable, we
did not even try to mention them all. A few of them which deal with rather
special classes of groups (e.g., the groups generated by reflections and the
braid groups) appear rather late in Chapter II.10. There and in Chapter II.9,
we briefly mention a fairly large number of topics which began to develop
before 1950. With one exception, this year is the cutoff date for our reports
on new lines of research. The exception is Chapter II.11 which deals with
the impact of mathematical logic on group theory. There, as well as in
Chapters II.2–II.8, we try to explain why we consider the subject matter
particularly important.

Unavoidably, the fact that we had to make choices when presenting the
developments after 1918 introduces the personal preferences of the authors
as an influence on the material appearing in this book. The same will be
true for our selection of references. Here, of course, improvements would
probably be possible, and we apologize for important omissions. However,
we did not feel any obligation to cover the literature after 1950 in any
systematic manner. The papers which appeared after this date are, in many
cases, quoted merely as illustrations.

The technical prerequisites for Part II are somewhat higher than those for
Part I. Although we give definitions of technical terms appearing in combinatorial group theory, we assume a knowledge of basic terms of algebra. We
have tried to keep the number and complexity of formulas to a minimum
and we show a certain preference for theorems which can be stated in a
simple manner.

Chapters II.12–II.14 are the analog of Chapter I.8. Here we are dealing
not with mathematics as such but with phenomena relevant to mathematical
research. What we are offering is not a systematic study but a number of
observations illustrated with examples from our special field.

Chapter II.2

Free Groups and Their Automorphisms

Much of this chapter will be devoted to an account of the results obtained by Jakob Nielsen (1890–1959), mainly in the years 1917–1924. Also, we shall use his papers to organize the material, giving, after each paper, an outline of later developments which were initiated by it.

The remark in Chapter II.1 about the disruptive effect of World War I can be illustrated by the publication dates of Nielsen's papers. We have mentioned Nielsen's Ph.D. thesis in Chapter I.6. It appeared in 1913. Nielsen, who was then a German citizen (although he was of Danish descent and became a Danish citizen after 1920 when his place of birth was returned to Denmark which had lost it in 1866) was drafted into the German navy in the same year. The war kept him there until 1918, and it was due only to special circumstances that he was able to publish two papers, dated 1917 and 1918, during this time. (For a detailed biography of Nielsen, see Fenchel [1960].) Even the physical appearance of the 1917 publication shows the influence of the war. The paper is now yellow and brittle, whereas in earlier volumes of the same periodical, the paper is still in very good condition. Also, the volumes of the *Mathematische Annalen* published during these years are much slimmer than those published earlier and later.

The title of Nielsen's paper of 1917 contains outdated terminology. Translated literally, it is: "The isomorphisms of the general infinite group with two generators." Today, we would write "The automorphisms of the free group on two free generators."

Nielsen himself explains the motivation for this paper. It is a continuation of the second part of his Ph.D. thesis (mentioned briefly in Chapter I.6 and cited as Nielsen [1913]). There Nielsen had investigated the fixed points of a topological self-mapping of a torus. The classes of topologically equivalent self-mappings form a group (the *mapping class group*) which, in this case, is isomorphic with the group of automorphisms of the fundamental group, the free abelian group of rank 2. NIELSEN [1917] shows that in the case of a torus from which a small disk has been removed, the mapping class group is isomorphic with the group of automorphism classes (i.e., the quotient group of the inner automorphisms in the automorphism group) of the free group F_2 of rank 2. Of course, F_2 is the fundamental group of the torus from which a disk has been removed. The importance of this paper is due not so much to its topological but to its group-theoretical contents. And it is the methods introduced by NIELSEN as well as his results which deserve attention.

The two main results obtained by NIELSEN [1917] can be formulated as follows (using only a slight modification of his own formulations):

Theorem $N1$. *Let ω be an automorphism of the free group F_2 on two free generators a, b. Let α, β be, respectively, the images of a, b under the action of ω. For any word W in a, b, let $s_a(W)$ and $s_b(W)$ be, respectively, the sum of the exponents of a and b in W. Then the matrix*

$$S = \begin{pmatrix} s_a(\alpha), & s_b(\alpha) \\ s_a(\beta), & s_b(\beta) \end{pmatrix}$$

has determinant ± 1, and any such matrix determines exactly one automorphism class (coset of the subgroup of inner automorphisms in the group of all automorphisms) of F_2.

Theorem $N2$. *The elements α, β of F_2 are images of a, b under the action of an automorphism of F_2 if and only if, in F_2*

$$\alpha\beta\alpha^{-1}\beta^{-1} = T(aba^{-1}b^{-1})^{\pm 1}T^{-1},$$

where T is an arbitrary element of F_2.

NIELSEN mentions in a footnote that Theorem N2 is due to MAX DEHN but that DEHN had given a different proof which, however, was never published and has not been found in the papers left by DEHN (which, in part, are undecipherable, being too sketchy or in shorthand).

NIELSEN's proof of Theorem N1 is based on a proof of the fact that the automorphism group of F_2 is generated by the obvious automorphisms which either exchange the generators or map one of them onto its inverse or onto the product with the other generator. To prove this fact, NIELSEN uses cancellation arguments. If a is a word $W(\alpha, \beta)$ in α, β, then, if α and/or β are words involving more than one symbol, a large number of cancellations must take place if we replace, in $W(\alpha, \beta)$, the elements α, β by their words in a, b. We mention here that NIELSEN considers the solution of the word problem in free groups as something obvious: If, in a word $W(a, b)$, we delete all subwords of the form aa^{-1}, $a^{-1}a$, bb^{-1}, $b^{-1}b$ and repeat this process again and again, we arrive, after finitely many steps, at a word $\overline{W}$ in which no such deletions are possible. $\overline{W}$ is called *freely reduced*, and the distinct freely reduced words are in one-to-one correspondence with the distinct elements of the free groups. (The first proof of this theorem was published by SCHREIER [1927a]. Of course, it holds, in a properly generalized form, for free groups with an arbitrary number of generators.

The proof of Theorem N2 uses the same type of cancellation arguments as the proof of Theorem N1. Of course, once one knows the generators of the automorphism group, the "if" part of Theorem N2 is obvious.

The next paper, NIELSEN [1918], establishes the fact that the automorphism group of the free group F_n on n free generators is again generated by the obvious automorphisms. Either permute the generators or keep all but one generator fixed and map the exceptional generator either onto its inverse or onto its product with another generator. Actually, this result had been pronounced already by VOGT [1889, pp. 17–20] and supported by an algebraic argument. But comparing the text of VOGT's paper with that of NIELSEN's makes it clear that VOGT's proof has a gap. Incidentally, VOGT remarks that STOUFF [1888] had already proved the same result using geometric arguments and the theory of discontinuous subgroups of $SL(2,\mathbb{C})$. Implicitly, FRICKE and KLEIN [1897 and 1912] use the same result. But NIELSEN's proof is definitely the first complete algebraic proof, the geometric proofs being of doubtful rigor.

It is worth mentioning that NIELSEN's 1918 paper no longer has a direct topological motivation. It is clearly of group-theoretical importance, generally speaking, since free groups are important (every group being a quotient group of a free group), and specifically because the automorphisms of F_n define for every n-generator group possible transitions from one set of generators to another. However, for $n > 2$, the full automorphism groups of F_n do not have the same significance for the topological self-mappings of a two-dimensional manifold with F_n as a fundamental group which they have in the case $n = 2$. There then exist automorphisms of F_n which cannot be induced by a topological self-mapping of the manifold.

NIELSEN's next group-theoretical paper appeared in 1921. (It is not his next paper but has the number 8 in his list of publications. The papers in between deal with problems in mechanics and with fixed points of topological self-mappings of two manifolds.) The title of this paper is, in English translation: *On calculating with noncommuting factors and its application in group theory*. (The paper is written in Danish.) Here, NIELSEN perfects the methods developed in his 1917/1918 papers for the purpose of investigating the finitely generated subgroups of *free groups* using this term for the first time in a published paper. (It had been coined by DEHN.) A brief and somewhat sketchy summary of the results of NIELSEN [1921] can be formulated as follows:

Let F_n be the free group on n free generators a_ν, $\nu = 1, 2, \ldots, n$. Let g_μ, $\mu = 1, \ldots, m$ be any m elements of F_n, given as freely reduced words in the a_ν. The g_μ generate a subgroup G of F_n. We can derive new generators of G by applying a finite number of times the transformations of the g_μ which NIELSEN had used to define the automorphisms of a free group, i.e., permuting the g_μ or replacing one of the g_μ by its inverse or by $g_\mu g_\rho$, where $\rho \neq \mu$. They are called *elementary Nielsen transformations*, a general Nielsen transformation being the composite of finitely many elementary ones. Let L be the total number of symbols $a_\mu^{\pm 1}$ appearing in the set of all of the g_μ. Then L is called the *length* of the set. NIELSEN gives a simple algorithm whose individual steps consist of elementary Nielsen transformations and subsequent free reductions. None of the steps increases the length of the set, and after finitely many steps, we arrive at a set of elements g_μ' which has minimal length and can be recognized because it satisfies certain conditions which are, in current terminology, described by saying that the set is *Nielsen reduced*. Some of the g_μ' may be the empty word 1. The remaining ones are shown by NIELSEN to be "independent", i.e., to be free generators of a free group.

These results have two consequences. One of them is the fact that a finitely generated subgroup of a free group is again a free group. DEHN, in a conversation with MAGNUS, mentioned that he had derived this result geometrically long ago, basing his proof on the fact that a connected subgraph of a tree (which is the graph of a free group) is again a tree. However, NIELSEN's paper certainly contains the first algebraic proof of this theorem, and also a very important algorithm.

Another consequence of NIELSEN's results is the fact that a finitely generated free group cannot be isomorphic with one of its proper quotient groups, i.e., a homomorphic mapping of F_n onto F_n always has a kernel of order 1. This has been observed by FEDERER and JÓNSSON [1950, p. 10]. Indeed, if the homomorphism η maps the a_ν onto elements g_ν of F_n, then the g_ν can, by a Nielsen transformation, be changed into a Nielsen reduced set. The number of elements in this set must again be n since, by abelianizing, it is easy to show that a group isomorphic with F_n cannot have fewer than n generators. On the other hand, NIELSEN had shown that the elements g_ν' of a Nielsen reduced set cannot satisfy any nontrivial relation, i.e., that a word $W(g')$ in the g_ν' can be equal to the empty word on F_n only if it can be changed into 1 by cancellations of subwords $g_\nu' g_\nu'^{-1}$ and $g_\nu'^{-1} g_\nu$. This proves that F_n cannot be isomorphic with any one of its proper quotient groups, a property which is definable for any group and is now called *hopfian*, after H. HOPF. The reason for this is the following.

In two papers, H. HOPF [1930 and 1931] had investigated the mappings of one two-dimensional manifold into another. These mappings also induced homomorphic mappings of the fundamental group of the original manifold onto that of its image, and HOPF had shown, with topological methods, that these groups indeed can not be isomorphic with any one of their proper quotient groups. The simplest case is the one where the fundamental group in question is free (and, of course, of finite rank), and it is somewhat surprising that for 30 years nobody (as it seems, not even NIELSEN himself) recognized that NIELSEN had given an algebraic proof (and a simpler one than the one given by HOPF with topological methods) for the fact that free groups of finite rank are hopfian, 10 years before HOPF had raised the question.

HOPF had, of course, recognized that his problem was of a general nature and in conversations had asked the question whether there exist finitely generated and, in particular, finitely presented groups which can be isomorphic with one of their proper quotient groups. (For infinitely generated groups, examples of this occurrence are very easy to find.) The question was passed on to MAGNUS by B. H. NEUMANN, and MAGNUS [1935] gave an algebraic proof very different from the one given by NIELSEN in the case of free groups. This seems to be the first place where HOPF's problem was stated explicitly in print. Some of the developments instigated by HOPF's question will be described in Chapters II.5 and II.10.

NIELSEN came back to the automorphisms of free groups with two papers (NIELSEN [1924a and 1924b]). The first of these, NIELSEN [1924a] is a very difficult paper. In it, NIELSEN derives finite presentations of the groups $A(F_n)$ of automorphisms of the free group F_n for $n \geq 3$. (The case $n = 2$ is easily settled and contained in NIELSEN [1917].) Since generators of $A(F_n)$ had been found earlier (NIELSEN [1918]), the problem is solved by producing a unique normal form for an arbitrary automorphism α as a word in these generators (where, however, α can be a word of arbitrarily great length) and by subsequently listing a set of relations which suffice to put every word into a normal form. An unsystematic poll taken by MAGNUS in 1970 seems to indicate that for about a decade after the death of NIELSEN in 1959 there existed no living mathematician who had read NIELSEN's paper in detail or would have been able to derive his result.

This situation changed dramatically with the appearance of three papers by McCOOL [1974, 1975a and 1975b]. In the first of these papers, McCOOL [1974], used a new method to derive a finite presentation for $A(F_n)$ which he then (McCOOL [1975a]) showed to be equivalent to the one found by NIELSEN. Finally, McCOOL [1975b] proved the following very general theorem:

Let $W_1, \ldots, W_m$ be any finite set of cyclically written words in the generators of F_n. Then the subgroup of $A(F_n)$ which fixes each of these words has a finite presentation and there exists an effective algorithm for constructing such a presentation.

It is interesting to note that MCCOOL uses methods developed by HIGGINS and LYNDON [1962 and 1974] which, in turn, deal with another problem arising from NIELSEN's work on automorphism groups of free groups and which was first solved by WHITEHEAD [1936a and 1936b]. Let $W_1, \ldots, W_t$ and $W_1', \ldots, W_t'$ be two finite sets of cyclically written words in the free generators of a free group F_n where $n < \infty$. Then WHITEHEAD gave a fully explicit algorithm which allows one to decide in a finite (and even predictable) number of steps whether the two sets can be carried into each other by an automorphism of F_n. WHITEHEAD's papers use difficult topological arguments. The paper by HIGGINS and LYNDON [1974] uses simpler ideas based on graph theory, and their arguments have been simplified further by HOARE [1979], who also showed how to simplify the proof of MCCOOL's theorem. A purely algebraic proof was given by RAPAPORT [1958]. We shall not try to describe the methods involved here or to give a complete list of references. Instead, we refer to LYNDON and SCHUPP [1977, pp. 21–49].

NIELSEN's paper [1924b] appeared almost simultaneously with his paper [1924a], but [1924b] is the later paper because it refers to the other one. The last two paragraphs clearly show that Nielsen [1924b] was motivated by questions about the structure of the automorphism groups $A(F_n)$ for $n > 2$, although the main topic of the paper is the derivation of a finite presentation for the group $GL(3, \mathbb{Z})$ of 3×3 matrices with integers as entries and with determinant ± 1. This group can be described as the group of automorphisms of a free abelian group of rank 3 or, as the title of NIELSEN's paper implies, as the group of affine transformations of euclidean three space which map a cubic lattice onto itself, keeping one particular point fixed. It is easily seen to be the quotient group $A(F_3)/K_3$ of $A(F_3)$, where K_3 denotes the subgroup of $A(F_3)$ consisting of those automorphisms which keep the cosets of the commutator subgroup F_3' of F_3 individually fixed. One of NIELSEN's results states that K_3 is generated by the conjugates (in $A(F_3)$) of a single automorphism which NIELSEN exhibits explicitly. He is able to do so by showing that $GL(3, \mathbb{Z})$ arises from $A(F_3)$ by adding one more defining relation to the presentation of $A(F_3)$ which NIELSEN had established in his paper [1924a].

The main part of NIELSEN [1924b] is devoted to the derivation of a finite presentation of $GL(3, \mathbb{Z})$. That such a presentation exists could have been derived immediately from the work of MINKOWSKI [1905]. There MINKOWSKI shows that $GL(3, \mathbb{Z})$ acts as a discontinuous group of affine transformations on a part of the euclidean space of six (real) dimensions which is defined by the coefficients of a positive-definite quadratic form in three variables and proves that $GL(3, \mathbb{Z})$ has a fundamental region which is a convex cone bounded by finitely many hyperplanes. In addition, MINKOWSKI showed that, apart from the apex of this cone, any point of its boundary can belong to the boundaries of only finitely many images of the cone obtained through the action of $GL(3,\mathbb{Z})$.

However, to use these arguments to produce an explicit presentation for $GL(3,\mathbb{Z})$ appears to be a formidable task and indeed NIELSEN chose another method without mentioning MINKOWSKI at all, although it is not unlikely that he knew MINKOWSKI's work. NIELSEN, like MINKOWSKI, uses a geometric language, but his proof is purely algebraic. Let M be any matrix in $GL(3,\mathbb{Z})$. Then he denotes the sum of the squares of the entries of M by $\sigma(M)$. The elements of the orthogonal subgroup O_3 of $GL(3,\mathbb{Z})$ are characterized by $\sigma(M) = 3$. NIELSEN uses a set of generators of O_3 as part of his set of generators of $GL(3,\mathbb{Z})$. As the remaining generators, he uses the transvections which are defined as matrices with entries $+1$ in the main diagonal and a single nonzero entry $+1$ outside. Next, he proves the following lemma:

Every element in $GL(3,\mathbb{Z})$ can be written in the form

$$\omega f_1 f_2 \cdots f_r, \qquad (*)$$

where $\omega \in O_3$ and f_ρ, $\rho = 1, \ldots, r$, are transvections or their inverses, and where, for $\rho = 1, \ldots, r-1$:

$$\sigma\left(f_\rho f_{\rho+1} \cdots f_r \right) > \sigma\left(f_{\rho+1} \cdots f_r \right). \qquad (**)$$

Now a set of relations for these generators will define $GL(3,\mathbb{Z})$ if it contains a set of defining relations for O_3 and in addition suffices to put the product of any element of type $(*)$ and a generator or its inverse again into the same shape (possibly changing the value of r but maintaining the inequalities $(**)$). The calculations needed for implementing this program are not easy and are demonstrated first in the much simpler case of $GL(2,\mathbb{Z})$. The resulting presentation is then simplified by NIELSEN.

We have given so many details of NIELSEN's paper because it illustrates an important method in a comparatively simple case. NIELSEN [1924b] did not deal with the case $GL(n,\mathbb{Z})$ for $n > 3$. However, it was shown by MAGNUS [1934a] that the case $n = 3$ is, in a sense, critical. It cannot be reduced directly to the case $n = 2$, although it is possible to reduce the cases where $n \geqslant 4$ directly in a very simple manner to the case where $n = 3$. It was found much later that the group $GL(2,\mathbb{Z})$ has, in fact, a structure which is very different from that of all of the $GL(n,\mathbb{Z})$ for $n \geqslant 3$. Whereas $GL(2,\mathbb{Z})$ has infinitely many normal subgroups which are both of infinite index and of infinite order, it has been shown by MENNICKE [1965] and by BASS, MILNOR, and SERRE [1967] that such subgroups do not exist in $GL(n,\mathbb{Z})$ for $n > 2$. Nevertheless, this result has nothing to do with group presentations. So far, we are not able to construct infinite groups in which every infinite normal subgroup is of finite index by starting with a suitable presentation unless we assume that there is an infinite normal abelian subgroup.

MAGNUS [1934] also confirmed that the kernel K_n of the natural homo-morphic mapping of $A(F_n)$ onto $GL(n, \mathbb{Z})$ is generated by the conjugates of a single element and, in addition, exhibited explicitly a finite set of genera-tors for K_n for all n. The next paper which provided information about the structure of $A(F_n)$ was a general theorem due to BAUMSLAG [1963] which states that the automorphism groups of finitely generated residually finite groups are themselves residually finite. This proves that $A(F_n)$ is residually finite. The same is true for the quotient group of $A(F_n)$ with respect to the group of inner automorphisms, according to GROSSMAN [1974]. The action of K_n on the augmentation ideal of F_n shows that K_n is residually torsion-free nilpotent. For this and a few related results, see MAGNUS [1980]. No finite-dimensional matrix representation for $A(F_n)$ is known for $n \geqslant 2$, and there are indications that none may exist; see MAGNUS and TRETKOFF [1980].

It is remarkable that a group about which there is, at least for $n > 2$, comparatively little known can be shown to be complete, i.e., having trivial center and no automorphisms except for the inner ones. This was proved for all $A(F_n)$ by DYER and FORMANECK [1975]. For the finite subgroups G of $A(F_n)$, DYER and SCOTT [1975] proved that the elements of F_n which are mapped onto themselves under the action of G form a free factor of F_n.

The papers published after 1934 dealing with $A(F_n)$ use a variety of methods and results, most of which were not available to NIELSEN in 1924. We cannot go into details here and have to refer to LYNDON and SCHUPP [1976] for information. But we would like to note that the history of the theory of $A(F_n)$ illustrates the fact that, even in the theory of infinite groups, there exist special groups defined for purely algebraic reasons which attract research work over long periods of time in spite of gaps of several decades in the sequence of papers that deal with them. In Chapter II.10 on mapping class groups, we shall encounter other special groups which have been the subject of research papers for even longer periods of time. But these groups have an immediate topological interest. Similarly, certain linear groups, in particular, the groups $SL(n, \mathbb{Z})$, are of immediate interest in number theory. But the groups $A(F_n)$ themselves (in contradistinction to some of their subgroups) are of importance mainly in connection with the Nielsen transformations which play an increasingly important role, e.g., in the theory of one relator groups. It seems that no topological space is known whose fundamental group is F_n and the topological self-mappings of which induce the full group of outer automorphisms of F_n if $n \geqslant 3$.

In his last paper, NIELSEN [1955] came back once more to the problem of finding free generators for a subgroup S of a free group F. If S is given by finitely many generating elements α_ν ($\nu = 1, \ldots, n$) which are words in the generators of F, then NIELSEN [1921] had shown how to construct explicitly a finite set of free generators of S. Now, in 1955, NIELSEN applies his method to the case where the set of α_ν is infinite. Of course, NIELSEN knew that the problem of proving that S is a free group even in this case had been solved by SCHREIER [1927] and he also cites later papers by LEVI [1930], HALL and RADO [1948], M. HALL [1949a], and FEDERER and JÓNSSON [1950], which deal with the same problem or with problems related to it. But obviously, NIELSEN's paper was motivated by the word problem. He notes that the kernel K of the homomorphic mapping of a free group F onto a group G which is defined by a presentation consists of a normal subgroup of F which is generated by the conjugates of the relators appearing in the presentation of G. The word problem for G is identical with the problem of deciding which elements of F belong to K. Now we quote NIELSEN (with a slight change of notation):

If K is determined by a given, infinite, related or nonrelated set

$$S: \alpha_1, \alpha_2, \ldots \tag{$*$}$$

of elements generating K, then there exists a function $L(l)$ such that all elements of K of length not exceeding l are contained in the subgroup K_L generated by the part $\alpha_1, \ldots, \alpha_L$ of ($*$); and for K_L a basis can be constructed in a finite number of steps. However, the dependence of L on l cannot, in general, be determined unless some further information can be gained from the structure of S. It is therefore only for finitely generated groups K that the identification of their elements can be assured unconditionally in a finite number of tests.

It is worth mentioning that the unsolvability of the word problem even for some finitely presented groups was proved by NOVIKOFF and by BOONE in the same year (1955) in which NIELSEN's paper appeared.

A less incisive logical aspect of the subgroup theorem for free groups is the fact that it is claimed to be true for all cardinalities of the order of the group and the subgroups. Most of combinatorial group theory deals with countable groups, although, as DEHN liked to point out in conversations, there is no reason why presentation should be confined to countable groups, and they are easily available in the case of some Lie groups. But LEVI [1930] was the first to point out that even in the case of the subgroup theorem for free groups with infinitely many generators, the proof that every subgroup is free requires a well ordering of the generators. It seems that not much attention has been paid to this question, but it is mentioned elsewhere, too, e.g., by FEDERER and JÓNSSON [1950], who assume that the subgroup is well ordered in a manner consistent with the lengths of the words representing its generators.

In principle, a large part of the problem of combinatorial group theory can be translated into problems about free groups. It certainly would not be expedient to do so. But the title of NIELSEN's 1921 paper, *Regning med ikke kommutative Faktorer* (Calculating with noncommutative factors) describes an exceedingly valuable technique which continues to be of great importance. In addition, NIELSEN's work may have contributed to a recognition of the importance of other "free" algebraic structures, e.g., associative rings and Lie rings.

We cannot give here a full evaluation of NIELSEN's work. It will appear prominently again in Chapter II.10. But apart from technical details, there is nothing which we could add to the obituary written by FENCHEL [1960].

Chapter II.3

The Reidemeister–Schreier Method

As indicated by its title, this chapter will deal mainly with a method rather than with theorems. Given a presentation of a group G, the Reidemeister–Schreier method shows how to compute a presentation for a subgroup H of G.

Both KURT REIDEMEISTER (1893–1971) and OTTO SCHREIER (1901–1929) published their first paper after World War I. Both of them made important contributions to combinatorial group theory; both of them were many-sided mathematicians; and both of them had advisors for their Ph.D. theses who were distinguished number theorists. Here, however, the similarity ends. REIDEMEISTER was, like DEHN, essentially a geometer. His influence on combinatorial group theory is largely that of a pioneer. His ideas were stimulating and had, at least in some cases, a long-lasting effect. SCHREIER, too, used geometric ideas, but he was above all a powerful algebraist who proved fundamental theorems. We shall try to substantiate these general remarks by giving a few specific data.

REIDEMEISTER's Ph.D. thesis deals with a topic in algebraic number theory (REIDEMEISTER [1921]) which had been proposed by E. HECKE. Of all the 71 papers listed in REIDEMEISTER's obituary by ARTZY [1972], this is the only one which deals with number theory. It rarely happens that a highly productive mathematician deserts the field of his Ph.D. thesis so consistently later on. Immediately after obtaining his Ph.D. degree, REIDEMEISTER started working on problems of differential geometry, obviously under the influence of BLASCHKE. Later, he became interested in topology, in particular, in knot theory. It is reasonable to assume that this interest was stimulated through WIRTINGER during the time (1922–1925) of REIDEMEISTER's stay at the University of Vienna. This assumption is supported by the fact that REIDEMEISTER [1926] mentions that the method of computing the fundamental group of a knot from its projection which appears in ARTIN [1925] goes back to WIRTINGER. REIDEMEISTER [1932a] mentions that the source was a talk WIRTINGER had given in 1905 but had never published. We shall later describe the details of REIDEMEISTER's contributions to combinatorial group theory which arose from his interest in topology. Apart from his papers, his books as well as his students have to be mentioned here also. His book on knot theory, which appeared in 1932, was the first and, until the appearance of CROWELL and FOX [1963], the only monograph on this subject. And his book on combinatorial topology, which appeared in the same year, contains a nearly up-to-date survey of combinatorial group theory. Of the papers instigated or influenced by him, those of PEIFFER [1949] and LYNDON [1950] are of particular importance for group theory.

For a full account of REIDEMEISTER's work, see ARTZY [1972].

SCHREIER's Ph.D. thesis had the title *Extensions of Groups* and appeared in two parts in 1926. His thesis advisor was FURTWÄNGLER, whose main field of interest was the classfield theory of algebraic number fields. This theory had been originated by HILBERT [1897] in his famous *Zahlbericht*. Since GALOIS, group theory had been closely related to the theory of algebraic number fields and HILBERT had extended this relationship to number-theoretical questions. We cannot say whether FURTWÄNGLER had a specific reason for being interested in group extensions already at that time, but this certainly was the case later when in 1930 he proved an important number-theoretical result by using group-theoretical methods involving group extensions. We shall describe this result later, in the chapter on metabelian groups.

Apart from FURTWÄNGLER, both WIRTINGER and REIDEMEISTER seem to have influenced SCHREIER. According to an unsigned obituary which appears (without page numbers but with a picture) as the first three pages following p. 106 of Volume 7 (1930) of *Abhandlungen aus dem Mathematischen Seminar der Hamburgischen Universität*, SCHREIER had attended courses given by these mathematicians during his student days in Vienna. This may explain his interest in combinatorial group theory which expresses itself in his first publication (SCHREIER [1924]). There he gives a simple algebraic proof of the fact that certain torus knots cannot be isotopic with their mirror images, generalizing a famous theorem of DEHN [1914]. Incidentally, the groups which appear here have the presentations

$$\langle A, B; A^\alpha B^\beta = 1 \rangle$$

and provide the simplest examples of free products with amalgamated subgroups—a topic which appears again in full generality in SCHREIER [1927a] and which will be described in Chapter II.4. Also, SCHREIER [1926a and 1926b (the Ph.D. Thesis)] will appear briefly in Chapter II.9. Apart from these papers which have a direct bearing on combinatorial group theory, we mention the fact that SCHREIER made important contributions to other parts of group theory. The classical Lie groups (finite dimensional, over the fields of real or complex numbers) can be considered as topological spaces. SCHREIER [1927b] showed that the fundamental group of such a space is always abelian. And SCHREIER [1928] found an important refinement of the fundamental Jordan–Hölder theorem, 39 years after the publication of HÖLDER's paper. It is rare that such a widely used and basic theorem can be deepened after such a long time. (In this case, something even more unusual happened. ZASSENHAUS [1934] discovered a second improvement of the theorem.)

Although SCHREIER used geometric ideas on various occasions, his work represents a strong trend towards algebraization and is in contrast with that of DEHN where at least the motivations if not the proofs were always geometric. In part, this was certainly in agreement with a general trend. Although REIDEMEISTER, like DEHN, was, above all, a geometer, his book on "combinatorial topology" (REIDEMEISTER, [1932b]) contains hardly any drawings. Abstraction and rigor were very much in fashion. This influenced the textbooks, too. One merely has to compare the earlier textbook by SCHÖNFLIES and DEHN [1930] on analytic geometry with the corresponding book by SCHREIER and SPERNER [1931 and 1935] (which was translated into English in 1951) on algebra and analytic geometry. There circulated an unpublished review of the latter book which said: "It is mainly a book on algebra. However, there are some applications to geometry, especially a proof of the fundamental theorem of algebra." For readers who studied mathematics after, say, 1950, we mention here that the term "fundamental theorem of algebra" had been coined by GAUSS (if not earlier) and was in circulation for about a century. It is the theorem which states that the field of complex numbers is algebraically closed.

We have dedicated much space to SCHREIER and his work. There can be no doubt that SCHREIER's paper of 1927, entitled *Die Untergruppen der freien Gruppen*, is one of the most important papers ever published on combinatorial group theory. It took a long time for all of its aspects to become effective, and it contains much more than the title indicates. We shall have to mention it again in several later chapters.

SCHREIER died in 1929 of what then was called a "general sepsis" at the age of 28. A few years later, the discovery of sulfur drugs by Domagk probably would have saved his life.

After our somewhat lengthy introduction, we shall now finally start with a discussion of the Reidemeister–Schreier method. It deals with the problem of finding a presentation for a subgroup H of a group G which itself is given by a presentation. Clearly, this is a meaningful problem only if we already know something about G. For instance, G may actually be trivial, i.e., of order 1, and have no proper subgroups at all, or G may be infinite but have no proper subgroups of finite index. (Simple and striking examples for these occurrences may be found in HIGMAN [1951a].)

The first case where a presentation of certain subgroups H of infinite index in a group G has been derived from a presentation of G is almost certainly due to DEHN, although he never published his result. However, it forms the basis for his approach to the investigation of one-relator groups (also unpublished but known to MAGNUS through conversations with DEHN), and a first indication of DEHN's result may be found in DEHN [1911], where he mentions that a certain subgroup of a free group on two generators is infinitely generated.

What DEHN found was this: Let G be a group with generators $a, b, c, \ldots$ and assume that the exponent sum of a in all relations is zero. Then G has a normal subgroup H whose elements are represented by all those words in the generators which have exponent sum zero in a. (The quotient group G/H is infinite cyclic, and the powers of a are coset representatives of H in G.) H then is generated by the elements

$$b_n = a^n b a^{-n}, \qquad c_n = a^n c a^{-n}, \ldots . \qquad n = 0, \pm 1, \pm 2, \ldots$$

We obtain defining relators for H as follows. Let $R(a, b, c, \ldots)$ be a defining relator of G. We can express it as a word in $b_n, c_n, \ldots$, using only free reductions and insertions and using $b_n, c_n, \ldots$, merely as abbreviations for words in $a, b, c, \ldots$. Assume that R then becomes a word

$$S(\ldots b_n \ldots; \ldots c_n \ldots; \ldots) \qquad\qquad (*)$$

where the dots indicate that more than one of the $b_n, c_n, \ldots$ may occur in S. Now all of the words

$$S(\ldots b_{n+t} \ldots; \ldots c_{n+t} \ldots; \ldots), \qquad t = 0, \pm 1, \pm 2, \ldots \qquad (**)$$

become defining relators of H, and we obtain a complete set of defining relators for H by applying this process to all defining relators R of G. Of course, the defining relator $(**)$ of H arises from R if we replace R by $a^t R a^{-t}$.

One may conjecture that DEHN's interest in this particular case derived from the fact that knot groups K can always be presented in such a manner that it is possible to apply the procedure described here and that, in the case of knot groups, the normal subgroup constructed is actually the commutator subgroup K'. By abelianizing K', one then obtains the important knot invariant which is known as the Alexander polynomial of the knot (see CROWELL and FOX [1963]). However, there are no known indications anywhere in the published or unpublished papers of DEHN which would support this conjecture. Actually, DEHN used this particular subgroup construction for his investigation of one-relator groups. We shall report on this in Chapter II.5. Nevertheless, it was knot groups which provided the motivation for a systematic study of subgroups of groups given by a presentation.

The fundamental paper by REIDEMEISTER [1926] bears the title, *Knots and Groups*. It starts with some expository remarks on group presentations and proceeds with the proof of the following results:

Let G be a group given by a presentation with finitely many generators g_i, $i = 1, 2, \ldots, n$. Let H be a normal subgroup of G and let v_j be a complete set of right coset representatives of H in G, written as words in the g_i. REIDEMEISTER considers only the case where j runs through a finite set of J values, where J is the index of H in G, but he observes that his construction will also work if J is infinite. The coset representative of H itself will always be the empty word 1. Furthermore, let

$$\overline{r_j g_i} = r_{j,i}$$

denote the particular representative of the coset of H which contains $r_j g_i$. (Of course, $r_{j,i}$ is itself one of the r_k, where k denotes a particular value of the subscript set of J symbols). Now the elements

$$h_{j,i} = r_j g_i \left(\overline{r_j g_i} \right)^{-1} = r_j g_i r_{j,i}^{-1}$$

generate H, and defining relations for H in terms of the $h_{j,i}$ can be obtained systematically by a well-defined process which, if both n and J are finite, requires only finitely many steps. We shall not describe this process here. It appears in all textbooks on combinatorial group theory, but it is somewhat complicated, particularly in the original version given by REIDEMEISTER which has been simplified by SCHREIER [1927a], HUREWITZ [1931], and later authors.

REIDEMEISTER's motivation for his paper becomes apparent in the second part. He first notes that, for all knot groups K, the quotient group K/K' of K with respect to its commutator group K' is infinite cyclic. Therefore, abelianizing knot groups does not enable us to distinguish different knots. However, for every integer $n \geqslant 1$, K has exactly one normal subgroup H_n of index n with cyclic quotient group K/H_n. Abelianizing H_n may now enable us to distinguish between different knots. This observation initiated a long series of papers by many authors who have shown how to distinguish between knots by using characteristic quotient groups of the knot groups. Only occasionally shall we mention some of these papers if they are of interest beyond their significance for knot theory. But there is another aspect of REIDEMEISTER's paper which deserves attention. He emphasizes the fact that the group H_n defined above is the fundamental group of an n-fold covering space of the space which has K as a fundamental group. Certainly, this is not a new insight. Probably, it is due to POINCARÉ; we have no reference and it is, of course, very difficult to claim that something is *not* due to POINCARÉ. But the priority question is irrelevant here. What matters is the following fact:

Topological interpretations of group-theoretical structures can provide new tools for proving group-theoretical theorems, and they may supply intuitive guidance for discovering or formulating them. The graph of a group was the first fruitful topological interpretation discovered. The use of covering spaces is the second one, and, later on, REIDEMEISTER (REIDEMEISTER [1932b]) began to develop it systematically. However, it took a long time before it became a generally recognized method in combinatorial group theory. Here we cannot go into later developments; they will be indicated in Chapter II.10. Another topological interpretation of group-theoretical structures was found by VAN KAMPEN [1933a and 1933b]. It, too, is now fully appreciated after many years during which it received mainly honorable mention, although it was never really forgotten. Again, we have to postpone a description of its effects until Chapter II.10.

SCHREIER [1927a] mentions REIDEMEISTER [1926]. Almost certainly he had attended a talk given by REIDEMEISTER in January of the same year at Hamburg. We are concerned here only with that part of SCHREIER's paper which deals with the computation of presentations for subgroups. Another part of the same paper will appear again, and in a fundamental role, in the next chapter.

SCHREIER first showed that the method introduced by REIDEMEISTER can be extended to cover both the cases where the subgroup H is not normal in G and also the cases where G is infinitely generated or where H is of infinite index. Next, SCHREIER showed that the system of defining relations can be simplified considerably if one chooses as representatives for H a system of words W in the generators of G which has the following property:

Every initial segment of any W (including the empty word 1) is again a coset representative.

A system of coset representatives satisfying this condition is now called a *Schreier system*. The existence of such systems was proved by SCHREIER using the following observation: If a word W, freely reduced in the generators of G, is of minimal length, i.e., if there can be no shorter word W' representing the same coset of H, then, automatically, all initial segments of W are also coset representatives of different cosets of H and all of them are of minimal length. A Schreier system in which all coset representatives satisfy this condition is called *minimal*, and a minimal Schreier system always exists (at least if G is countable).

The use of these special systems of coset representatives made it possible for SCHREIER to prove the following theorem.

If F is a free group of any (finite or infinite) rank, then every subgroup of F is free. If F is of finite rank n and if H is a subgroup of finite index j in F, then the rank of H is exactly

$$1 + j(n-1).$$

If H is a normal nontrivial subgroup of F and j is infinite, then H is always of infinite rank.

SCHREIER gives two proofs of this theorem. One of them is purely algebraic; the other one uses the "coset graph" of a subgroup H in a group G given by a presentation. This is a natural generalization of the graph of a group. The points of the graph represent cosets of H rather than elements of G. Finally, SCHREIER gives a strikingly simple solution for the word problem in a free group, showing that a freely reduced nonempty word in the generators cannot represent the unit element. His proof is so simple that we can reproduce it here:

Let

$$W = S_1 S_2 \ldots S_r$$

be a word in the generators of the free group F, where the S_i $(i = 1, \ldots, r)$ denote either a generator or the inverse of a generator but where it never happens that S_i and S_{i+1} represent a generator and its inverse. We map F into a finite permutation group as follows: A generator not appearing in W is mapped onto the identical permutation of $r + 1$ symbols. The other generators are mapped onto elements of the symmetric group of permutations of $r + 1$ symbols by postulating that the generator appearing in S_i (either itself or as its inverse) maps the symbol i onto $i + 1$. These data do not determine the permutations assigned to the generators in a unique way, but they are consistent since S_i and S_{i+1} are never inverses of each other. And no matter how we complete the definition of our permutations, they have the property that the product $S_1 S_2 \ldots S_r$ maps the symbol 1 onto the symbol $r + 1$. Therefore, W cannot represent the identity in F. In fact, SCHREIER herewith proved that F is residually finite (a term introduced about 30 years later by P. HALL), i.e., that for every element $g \neq 1$ in F, there exists a finite homomorphic image of F in which the image of g is again $\neq 1$.

The proofs of several of SCHREIER's results have been streamlined by LEVI [1930] and HUREWITZ [1931]. LEVI also notes that, at least implicitly, SCHREIER had assumed that the generators of a free group must be well ordered and that this assumption is indispensable if one wishes to apply the Reidemeister–Schreier method to groups with an infinite number of generators. Nowhere in the literature before 1955 have we found a reference to the difficulties which may arise in the case where a group is given by a presentation with infinitely many (even countably infinite) defining relators.

The terms "recursively enumerable" and "recursive," which originated in mathematical logic and which now play an important role in combinatorial group theory, did not enter the field until the time of the publications of NOVIKOV and BOONE. Some information about these questions will be given in Chapter II.11.

KARRASS and SOLITAR, in Sections 2.4 and 3.2 of MAGNUS, KARRASS, and SOLITAR [1966] sharpened two of SCHREIER's results. They showed that any subgroup H of infinite index in a free group F is infinitely generated if it contains a nontrivial normal subgroup of F and that the Reidemeister–Schreier method produces a Nielsen-reduced set of generators for a subgroup H of a free group F if and only if the system of coset representatives of H in F is a minimal Schreier system.

The theory of subgroups of free groups is far from being restricted to results of the type mentioned here. In particular, sequences, especially ascending and descending sequences of subgroups of free groups, have been investigated for various purposes and from various points of view. But although many of the results are obtained through methods established by NIELSEN or SCHREIER, they belong to a later level of development of the theory, and we have to refer the reader rather summarily to MAGNUS, KARRASS, and SOLITAR [1966] and LYNDON and SCHUPP [1976] for details.

The application of the Reidemeister–Schreier method to the theory of free groups is, in some respect, particularly noteworthy since it leads to the proof of a general theorem. There are a few other consequences of a general but rather obvious nature, for instance, the fact that a subgroup H of finite index in a finitely presented group G is itself finitely presented and contains a subgroup N which is normal and of finite index in G. Also, it follows that a finitely generated group can have only finitely many subgroups of a given finite index, and that their intersection is again of finite index and also characteristic in the whole group. But the importance of the Reidemeister–Schreier method is based on the fact that it clearly establishes both the data needed for the computation of a presentation of a subgroup and the process to be used in carrying out this computation. In many particular cases, it is easy to provide the necessary data, and the Reidemeister–Schreier method has been used in numerous papers for the investigation of individual groups given explicitly by a presentation. The case studied by DEHN and mentioned earlier appears most often in the literature.

The following problems are closely associated with the problems from which the Reidemeister–Schreier method arises.

Problem 1. *Let G be a group given by a presentation. Let S be a set of words in the generators of G. Then the elements of S generate a subgroup H of G. Find a presentation for H.*

Problem 2. *Let G be a group given by a presentation. Let R be a set of distinct elements of G which includes the unit element. Find all subgroups H of G which have R as a complete set of right coset representatives. (Of course, one will, in general, restrict R to a set of words which have the properties of a Schreier system).*

There exist substantial contributions to Problem 1, some of which will be discussed in the next chapter in a different context. Right now, we shall use these problems merely as headlines under which we can accommodate a few scattered but interesting papers.

If G is a finite group, TODD and COXETER [1936] developed a practical method to solve Problem 1, and COXETER and MOSER [1972, p. 16] note that this method is sufficiently mechanical to admit the use of an electronic computer. They also mention several papers in which the method has been programmed for automatic execution. But these results as well as the extensive use of electronic computers in the theory of simple groups of finite order are beyond the scope of the present monograph.

Problem 2 has hardly been touched at all. However, a paper by B. H. NEUMANN [1933] has to be mentioned here for two reasons. One of them is its motivation. The problem it deals with arose in the foundations of geometry. It was formulated by ARNOLD SCHMIDT, HILBERT's last Ph.D. student, and was, according to an account given by B. H. NEUMANN, transmitted to him via BERNAYS, H. HOPF, and I. SCHUR. (For the geometric aspect, see A. SCHMIDT [1934].) The question, as formulated by BERNAYS, is the following one: Let $G = SL(2, \mathbb{Z})$ denote the unimodular linear group of degree 2 over the integers. Does G contain a subgroup H with the following properties?

(i) Every ordered pair a, b of coprime integers a and b appears exactly once as the first column in an element of H.

(ii) H contains the matrix

$$A = \begin{pmatrix} 0 & -1 \\ 1 & 0 \end{pmatrix}.$$

NEUMANN showed that this is equivalent to finding all subgroups H which contain A and have all the powers of

$$T = \begin{pmatrix} 1 & 1 \\ 0 & 1 \end{pmatrix}$$

as coset representatives in G. Using a presentation of G, NEUMANN then proved that the possible subgroups H are in one-to-one correspondence to the solution of the following functional equations for one-to-one mappings f

of the integers onto themselves:

$$f(f(n)) = n, f(n-1) = 1 + f(f(n)+1), f(0) = 0; \qquad n \in \mathbb{Z}.$$

Finally, NEUMANN explicitly constructed a set of such mappings which has the cardinality of the continuum.

In spite of its special nature, we have given a rather detailed account of this paper for the following reasons.

(i)   As far as we know, this is the only paper on combinatorial group theory which is directly motivated by a specific problem in the foundations of geometry. Although B. H. NEUMANN stated in interviews that his approach to mathematics has been strongly influenced by his interest in HILBERT's work in this field, the effects of this influence appear to be of a more indirect nature, expressing themselves in the type of problems investigated rather than in their direct applicability.

(ii)   The paper is the first (and also a very nontrivial) application of the Reidemeister–Schreier method which establishes the existence of certain subgroups of infinite index. In most applications, the existence of the subgroup for which a presentation is to be computed is known beforehand.

(iii)   The group $SL(2,\mathbb{Z})$ is one of the most thoroughly investigated countably infinite groups, since it plays an important role both in number theory and in the theory of automorphic functions.

The paper appeared in the *Proceedings of the Prussian Academy* and is now not easily accessible in most places. A short survey of it appeared in MAGNUS [1974a] and it is the basis of papers by MAGNUS [1973] and C. TRETKOFF [1975].

Some aspects of a paper by P. HALL [1936] also are related to Problem 2. Let G be a finite group, and let $g_n(G)$ be the number of ordered n-tuples $c_1,\dots,c_n$ of elements of G which generate G. HALL calls $g_n(G)$ the nth *eulerian function of* G. If G cannot be generated by n elements, $g_n(G) = 0$. Otherwise, G is a quotient group F_n/N of a free group F_n of rank n, and $g_n(G)$ is the number of ordered n-tuples of coset representatives of N in F_n which, together with the elements of N, generate F_n. More generally, we may postulate that only such n-tuples of generators shall be admitted which satisfy certain relations. Then F_n is replaced by an n-generator group which has these relations as defining relations.

We cannot go into the details of HALL's paper which produces a wealth of surprising results. For instance, if G is a simple group of composite order,

HALL's method provides a remarkably simple method to compute the largest number e such that the cartesian power G^e can be generated by n elements. This result has been used by STORK [1971] to construct a characteristic subgroup of F_2 which is of finite index but not fully invariant and whose quotient group in F_2 is not solvable.

HALL mentions that part of his paper had been anticipated by L. WEISNER [1935]. Its main impact seems to have been on general combinatorics rather than on group theory (see, e.g., ROTA [1964]).

Chapter II.4

Free Products and Free Products with Amalgamations

In this chapter, we shall describe the emergence of two new ideas which had
a profound influence on the development of combinatorial group theory.
Their creation is associated with three names: O. SCHREIER, who appeared
prominently in Chapter II.3, E. ARTIN (1898–1962), AND A. G. KUROSH
(1908–1971).

ARTIN was a very many-sided mathematician. He appears in combinatorial group theory only as a sporadic guest, but with highly original
contributions. We shall refer to him again in Chapter II.6 in connection
with a group-theoretical problem which arose in his work on class field
theory and in Chapter II.10 in connection with his work in topology. He is a
prominent figure in Twentieth Century mathematics, and he was influential
not only through his ideas and his work, but also through his books and his
lectures. Nevertheless, we have to abstain here from saying more about him
since we cannot do justice to his work without going too far away from the
topic of the present book.

KUROSH, too, worked in several fields, but he is a central figure in
combinatorial group theory. This is due not only to his important contributions to the field, one of which is now permanently attached to his name,
but also through his influence on other mathematicians and his comprehensive monograph on group theory, the first edition of which appeared in
1944. Other parts of group theory had flourished in Russia well before his
time, but there can be no doubt that he and his students have been
instrumental in making the Soviet Union a prominent center of research in
combinatorial group theory.

KUROSH started as a topologist. His teacher was P. S. ALEXANDROV at
Moscow University where KUROSH obtained his first Ph.D. in 1930. In the
same year, KUROSH came in contact with O. JU. SCHMIDT and was admitted
to SCHMIDT's group theory seminar in Moscow. A prerequisite for admission was that one had to have results of one's own in this field. We do not

know the details, but it seems plausible to assume that KUROSH's first paper which appeared in 1934 was influenced in a general way by the ideas of O. JU. SCHMIDT (1891–1956), who was probably the first group theorist who tried to include infinite groups in a systematic account of abstract group theory. Until the publication of his book in 1916, the textbooks treated infinite groups as groups of transformations of a metric space. We shall explain later, after the description of KUROSH's results on free products, the relation between these and the results of SCHMIDT on direct products. According to KUROSH's obituary of SCHMIDT (which appeared in Russian in *Progress of Mathematical Sciences*, vol. 11, Issue 6 (72), November-December 1956), he had an outstanding influence as an organizer of group-theoretical research in the Soviet Union and as founder of the Moscow school of algebra. But his activities went far beyond mathematics. He organized expeditions to the Arctic in which he took part himself. According to KUROSH, these enterprises were motivated by his deep-rooted interest in cosmogeny, a field to which he devoted the better part of 15 years of his life. We do not know which arguments he, a mathematician, used in order to persuade the Soviet Government in 1930 to provide the equipment for his arctic expeditions.

Starting now with a description of the mathematical results involved, we have to mention first a section in a book by F. KLEIN [1926, pp. 361–364] which came out in a new edition shortly after the death of the author and contains a "third part" which had not been written by him, but had been arranged by the editor, W. BLASCHKE. In a short preface, BLASCHKE accredits the section on free products to ARTIN. However, SCHREIER [1927] mentions, and ARTIN [1926] confirms, that the results presented there had been found independently both by ARTIN and himself. We quote (in translation) the opening statement of the essay by ARTIN [1926]:

Consider a finite or infinite set of groups $G_1, G_2, \ldots$. We admit the occurrence of isomorphic groups in this set but we postulate that there is no rule of composition for any two elements from distinct groups (whether they are isomorphic or not).

We define now the free product G of our groups as the set of the following symbols.

1. The unit 1.
2. All symbols of the form

$$A = a_1 a_2 \cdots a_n,$$

where a_i denotes any element of our groups $G_1, G_2, \ldots$.

But here the following conditions have to be satisfied:

(a)   No a_i may be equal to 1.
(b)   Adjacent elements a_i and a_{i+1} must belong to different groups.

Two elements of G are called equal:

$$a_1 a_2 \cdots a_n = b_1 b_2 \cdots b_m$$

if and only if $n = m$ and if $a_i = b_i$ for all i.

ARTIN then proceeds to show that, using a suitable (and group-theoretically obvious) rule of composition for elements, G indeed becomes a group. ARTIN calls the expressions defined above the "normal form of the group elements."

Strangely enough, ARTIN concludes his general discussion with (quoted in translation):

> Now the meaning of our discussions becomes visible. In the same manner in which the consistency of the rules for calculations with complex numbers can be derived from a proof of the validity of the elementary rules of calculation for pairs of numbers, we have now recognized the consistency of the rules for noncommutative calculations. The importance of this insight for the foundations of group theory is evident.

In a footnote, ARTIN adds:

> In the special case where all factors of the free product are infinite cyclic groups, M. Dehn has proved its consistency geometrically by constructing the socalled *Gruppenbild*.

Finally, as a sort of afterthought, ARTIN explains that the free product of a cyclic group of order 3 and a cyclic group of order 2 could also be defined as a group of two generators b, c with defining relations $b^3 = 1$, $c^2 = 1$.

ARTIN had given here, for the first time, a purely algebraic solution of the word problem for free groups which can be defined as free products of infinite cyclic groups. (The second very elegant proof by SCHREIER we have already mentioned in Chapter II.3.) But the true importance of the free product for the theory of groups is expressed much more accurately in the opening sentence of the paper by KUROSH [1933]:

> The group G is called the *free product* of its subgroups H_i, $G = \prod H_i$ (where i runs through an arbitrary set of subscripts) if every element $g \neq 1$ of G can be represented uniquely as a product
>
> $$g = h_1 h_2 \cdots h_n; \qquad h_i \neq 1, h_i \in H_i, H_i \neq H_{i+1} \tag{1}$$
>
> (where, however, $H_i = H_j$ is admissible if $j \neq i + 1$). Therefore, there exists no nonidentical relation in G which connects elements of different H_i.
>
> Thus, every free product may be considered as being definable through the totality of generators and defining relations of all of the factors H_i.

In the language used by DEHN, one would say that the normal form (1) for the elements of G provides an explicit solution for the word problems of G if the word problems in all of the groups H_i have been solved.

KUROSH [1933] proved restricted versions of the theorems below which were fully proved by KUROSH [1934]. They are, in the version given by him:

I. The Kurosh Subgroup Theorem. *Every subgroup F of G can itself be decomposed into a free product*

$$F = {}_\beta\Pi F_\beta,$$

where every factor F_β is either an infinite cyclic group or conjugate (in G) with a subgroup of one of the groups H_i. (F may consist of only one factor.)

II. The Isomorphism Theorem. *If G can be represented in two different ways as a free product of freely indecomposable subgroups:*

$$G = {}_i\Pi H_i = {}_\beta\Pi F_\beta,$$

then there exists a one-to-one correspondence between the groups H_i and F_β such that corresponding groups are isomorphic. If corresponding groups are not infinite cyclic, then they are conjugate in G.

The historical importance of these theorems is evident. They are the first general structure theorems in combinatorial group theory, and they have found many applications later on. Looking backward to the first appearance of the free product in F. KLEIN's "composition of groups" in 1883 (see Chapter I.3), we now see that (one is tempted to say, "finally,") algebra had caught up with geometry, at least to some extent. KLEIN would have seen, without difficulty, from the shape of its fundamental region, that in a group with two generators b, c and defining relations

$$b^m = c^n = 1,$$

the only elements of finite order are conjugates of powers of b or c. Now this is an obvious consequence of the Kurosh subgroup theorem. In fairness to the power of geometric methods, we have to note here that the so-called "triangle groups" with the presentation

$$b^m = c^n = (bc)^k = 1$$

still pose now, 50 years later, considerable difficulties to a purely algebraic proof of the fact that elements of finite order are conjugates of powers of b, c or bc, at least if m, n, k are not sufficiently large (see MAGNUS [1974a, pp. 100–101 for references].

The isomorphism theorem was supplemented by BAER and LEVI [1936] with the following result.

III. Refinement Theorem. *Given any two decompositions of a group G into a free product, there exists a third decomposition of G into a free product such* *that each factor of the first two decompositions is itself a free product of factors of the third decomposition.*

Next, KUROSH [1937] showed that there are groups which do not admit any decomposition into a free product of freely indecomposable groups. He gave the following group T as an example:

Generators:

$$a_0, a_1, a_2, \ldots, a_n, \ldots,$$

$$b_1, b_2, \ldots, b_n, \ldots$$

Defining relations:

$$a_n b_n a_n^{-1} b_n^{-1} = a_{n-1},$$

$$n = 1, 2, 3, \ldots.$$

The group T is the limit of an ascending chain of free groups or, as we would say today, it is *locally free*, i.e., every finitely generated subgroup of T is a free group. Now T is the free product of the infinite cyclic group generated by b_1 and the subgroup T' of T generated by

$$a_1, a_2, \ldots a_n, \ldots,$$

$$b_2, \ldots b_n, \ldots$$

which is obviously isomorphic with T. Therefore, T could be, at best, the free product of infinitely many infinite cyclic groups. But KUROSH shows that this is not true, proving that T cannot, in any manner, decompose into the free product of infinitely many groups. For this proof, he needs the following interesting lemma.

If a group G is the free product of two subgroups H_1 and H_2, and if the commutator

$$g_1 g_2 g_1^{-1} g_2^{-1}$$

of two elements g_1, g_2 of G belongs to H_1 and is $\neq 1$, then both g_1 and g_2 belong to H_1.

The next step in the development of the theory of free products is the following:

Gruschko–Neumann Theorem. *Let G be a finitely presented group which is isomorphic with the free product of two groups B and C. Then there exists a Nielsen transformation of the generators g_i $(i = 1, \ldots, n)$ of G which maps them onto two sets of generators b_j $(j = 1, \ldots, r)$ and c_k $(k = 1, \ldots, s)$ such that, in* *terms of these generators, the defining relators of G can be written as words in b_j alone or in c_k alone.*

Here $r + s \leq n$, and the following corollary holds.

The minimal number of generators of a free product equals the sum of the minimal numbers of generators for the factors.

Although the papers by GRUSCHKO [1940] and B. H. NEUMANN [1943b] are 3 years apart, there can be no doubt that their results were obtained independently. World War II inhibited the international exchange of literature to a great extent.

Most of the later papers dealing with properties of free products or of free products of special types of groups are beyond the range of our account which does not have room for too many details or special results. But the theory of free products also appears in a broader algebraic context. We shall combine its description with a brief analysis of some characteristics of KUROSH's work.

In his first algebraic paper, KUROSH [1932] contributed to the analog of the isomorphism theorem, formulated above, for the case of direct rather than free products. The situation is much more complicated here than in the case of the free product, at least if the group G has a nontrivial center, and it is dealt with in one of the most important papers by O. SCHMIDT (i.e., O. JU. SCHMIDT) [1928]. Clearly, the paper by KUROSH [1932] was written under the influence of SCHMIDT, and it is highly plausible that later KUROSH may have studied the free product as an analog of the direct product. Both the direct and the free product are binary compositions which are defined for any groups as factors: the composition is associative and commutative and (something KUROSH may have remembered from his previous study of topology), both of them have a topological meaning since there exist products of topological spaces such that the fundamental group of the product space is either the free or the direct product of the fundamental groups of the factors. The question whether there exist other universally defined products of groups with some of the properties of direct and of free products has been taken up by KUROSH's student GOLOVIN [1950], who constructed an infinite sequence of such products the first of which had been discovered earlier by LEVI [1944]. GOLOVIN's first paper was followed by several important papers by him and other authors on the same subject. However, we abstain from giving any details for two reasons: The developments until 1964 relating to this topic have been summarized by MAGNUS, KARRASS, and SOLITAR [1966, Section 6.4], and since then not much work has been done in this direction.

In spite of the importance of Kurosh's work for the development of combinatorial group theory, only a small part of it belongs to this field if we define it narrowly as the theory of group presentations. However, KUROSH also did very much to establish a theory of infinite groups (outside of the theory of Lie groups) which is not based on group presentations.

The paper by DIETZMAN, KUROSH, and UTZKOW [1938] on Sylow subgroups of infinite groups provides a good illustration of this type of work. His monograph *Teoriya Grupp* (KUROSH [1944]) and, even more, the expanded English version of it (KUROSH [1955 and 1956]), contains systematic studies of problems arising in a general theory of infinite groups. We cannot go into details, but we mention at least one feature, namely, the investigation of groups with prescribed *local properties*. Here, a group is said to have a particular local property like being solvable, finite, free, etc. if every finitely generated subgroup has this property.

Beyond his interest in a general theory of infinite groups, KUROSH saw group theory as a part of a greater whole. His views are stated explicitly and carefully in the introduction of his book *Lectures on General Algebra* which appeared in English translation in 1963. It should be noted that in spite of its emphasis on the importance of general concepts, the book contains a section on something as special as the quaternion algebra and the Cayley algebra.

In the bibliography of KUROSH's book on the theory of groups, the author with the greatest number of references is REINHOLD BAER (1902–1979). This is no coincidence. Although BAER made only sporadic contributions to combinatorial group theory, he is a very important author for nearly all aspects of the theory of infinite or finite groups and for numerous problems in other parts of algebra. All of this makes BAER very congenial with KUROSH. As it happens, the similarities between these mathematicians go even further. BAER, too, started as a topologist (BAER [1927 and 1929]) and his Ph.D. advisor was H. KNESER, like P. S. ALEXANDROV, a leading topologist of his time. Like KUROSH, BAER had many students who not only wrote a Ph.D. thesis under his guidance but continued to be active in mathematical research. And, like KUROSH, he was definitely an algebraist. His book *Linear Algebra and Projective Geometry* (BAER [1952]) is, like the much more elementary book by SCHREIER and SPERNER [1931 and 1935] almost pure algebra. As an aside, we mention here that BAER's coauthor FRIEDRICH LEVI in their joint paper on subgroups of free products was, like DEHN and REIDEMEISTER, also a geometer who wrote a book on geometric configurations (LEVI [1929]).

We now turn to a discussion of the most important aspect of the paper by SCHREIER [1927a]. It consists of a generalization of the concept of a free product. SCHREIER called it the *free product with amalgamated subgroups* and it is also known as a *generalized free product*. The section of SCHREIER's paper which introduces this concept has the heading *An existence theorem* and takes up only three pages of his 22-page paper. We start with a translation of SCHREIER's definition:

> Let M be a set of groups G_i, where i runs through an arbitrary indexing set, and let H be an additional group. Assume that each one of the groups G_i contains a subgroup H_i isomorphic with H. If h denotes an arbitrary element of H, let h_i denote the corresponding element of H_i as defined by a fixed isomorphism between these groups. We shall try to construct a group P with the following properties:
>
> 1. For every G_i (with elements g_i) in M, P contains a subgroup $\overline{G}_i$ (with elements $\overline{g}_i$) isomorphic with G_i. Let $\{g_i \leftrightarrow \overline{g}_i\}$ define (for all i) an isomorphism of G_i with $\overline{G}_i$. Under these isomorphisms, each element h_i of H_i shall correspond to the element $\overline{h}_i$ of $\overline{H}_i$ in $\overline{G}_i$.
> 2. The subgroups $\overline{G}_i$ generate P, that is, every element of P can be composed of elements of the subgroups $\overline{G}_i$.
> 3. The subgroups of P which, according to the isomorphisms $\{\leftrightarrow\}$ correspond to the subgroups $\overline{H}_i$ of $\overline{G}_i$ are identical. This means that P contains a subgroup $\overline{H}$ with elements $\overline{h}$ which is isomorphic with H by means of an isomorphism $\{h \leftrightarrow \overline{h}\}$ and we have $\overline{h}_i = \overline{h}$ for all i.
> 4. The group P is—in a sense to be made precise below—as general as possible.

The term "as general as possible" is explained as follows: Clearly, every element of P can be put into one of the following forms:

$$\overline{h} \qquad \text{or} \qquad \overline{h}\,\overline{a}_1\overline{a}_2 \cdots \overline{a}_n, \qquad\qquad (*)$$

where $\overline{h}$ is an element of $\overline{H}$, $n \geq 1$, and $\overline{a}_1,\ldots,\overline{a}_n$ belong to certain groups $\overline{G}_{i_1},\ldots,\overline{G}_{i_n}$ and are fixed right coset representatives of cosets $\neq \overline{H}$ of $\overline{H}$ in $\overline{G}_{i_1},\ldots,\overline{G}_{i_n}$. Finally, $i_j \neq i_{j+1}$ for $j = 1,2,\ldots,n-1$. Now the "existence theorem" states that there exists a group P in which the form $(*)$ is unique for every element. This is the group P which is "as general as possible," and $(*)$ is the normal form of its elements.

In the introduction to his paper, SCHREIER [1927a] mentions that this result allows us to solve the word problem for groups with two disjoint sets X and Y of generators and a single defining relation

$$f(X)g(Y) = 1,$$

where $f(X)$ and $g(Y)$ are freely reduced nonempty words in the generators occurring in X and in Y, respectively. Clearly, this represents a generalization of his result in SCHREIER [1924]. The 1924 paper has the title: *On the groups $A^aB^b = 1$*, and apparently provided the starting point for his later investigation.

Due to its more general nature, the free product with amalgamations leads to subgroup theorems which are much more involved than those for the ordinary free product. The literature is rather extensive. B. H. NEUMANN [1954] gives a detailed account of everything that was known at that time. But this is by no means the end of the story. For additional references, refer to LYNDON and SCHUPP [1976].

The structure theorems are only a small part of the great multitude of results which have been obtained by using the concept and properties of a free product with amalgamations. One type of these results consists, naturally, of the solution of word problems. We start our account by giving a rather special but instructive example which is due to G. HIGMAN [1951a].

Let H_3 denote the group with generation a, b, c and defining relations

$$aba^{-1} = b^2, \qquad bcb^{-1} = c^2, \qquad cac^{-1} = a^2$$

and define the group H_4 by four generators a, b, c, d and the relations

$$aba^{-1} = b^2, \qquad bcb^{-1} = c^2, \qquad cdc^{-1} = d^2, \qquad dad^{-1} = a^2.$$

Then HIGMAN observes that H_3 is the trivial group of order 1. (Actually, to show this requires some clever calculations). Next, HIGMAN proves that H_4 has no subgroups of finite index by showing that the trivial group is the only finite quotient group of H_4. (For finitely generated groups, these statements are equivalent). Now the question arises: Is H_4, too, the trivial group? If not, then there clearly exists a finitely generated infinite simple group, something which up to that time was not known to exist. (It was, of course, known that, e.g., the quotient groups of the groups of $n \times n$ matrices, with determinant $+1$ and entries from an infinite field, with respect to their center are simple for $n \geq 3$. But all of these groups are infinitely generated). Now HIGMAN observed that H_4 is indeed infinite. Consider first the group H_1 defined by

$$H_1 = \left\langle a, b; aba^{-1} = b^2 \right\rangle.$$

According to the general theory of one-relator groups, both a and b generate infinite cyclic subgroups in H_1. Therefore, the group H_{12} defined by

$$H_{12} = \left\langle a, b, c; aba^{-1} = b^2, bcb^{-1} = c^2 \right\rangle$$

is a free product of H_1 with another one-relator group H_2 (generated by b and c) with an infinite cyclic subgroup (generated by b) amalgamated. From the normal form for elements in H_{12}, it follows that a and c generate a free subgroup of rank 2 in H_{12}. The same is true for the subgroup generated by a and c in the group H_{34} defined by

$$H_{34} = \left\langle c, d, a; cdc^{-1} = d^2, dad^{-1} = a^2 \right\rangle.$$

So, finally, H_4 is the free product of H_{12} and H_{34} with free amalgamated subgroup of rank 2, generated by a and c. This proves that H_4 is infinite.

HIGMAN had to leave open the question whether H_4 might itself be a simple group or not. It was proved later by SCHUPP [1971] that this is not the case at all. Using a term coined by HIGMAN, SCHUPP proved that H_4 is *SQ-universal* which means that every countable group is isomorphic with a subgroup of a quotient group of H_4. Since B. H. Neumann [1937b] had already proved that there exists a noncountable set of nonisomorphic two-generator groups, it follows that H_4 has in fact a noncountable number of distinct normal subgroups. We note here that this is true not only for individual other groups but for large classes of them as well, e.g., for all free products of two nontrivial groups, not both of which are of order 2. For the literature on this topic, see Chapter V.10 of Lyndon and Schupp [1976]. *SQ*-universality provides a model example of the introduction of a concept which leads to productive research.

We shall encounter other and less special applications of the theory of free products with amalgamations later on, particularly in the next chapter on one-relator groups. But by far the most important consequence of Schreier's theorem is undoubtedly the theory of *HNN extensions*. Although this theory was introduced by using the theory of the free product with amalgamations, and although both theories can be shown to be equivalent, the applications of HNN extensions are, if not more numerous, then at least more spectacular than those of the generalized free product. This fact appears to be the reason why Lyndon and Schupp [1976, pp. 178–187] introduce the HNN extension first.

We now quote from the paper by Higman, Neumann, and Neumann [1949]:

> Let μ be an isomorphism of a subgroup A of a group G onto a second subgroup B of G (B need not be distinct from A). Then there exists a group H containing G, and an element t of H, such that the transform of t by any element of A is its image under μ:
>
> $$t^{-1}at = \mu(a) \qquad \text{for all} \quad a \in A.$$
>
> If G is locally infinite, then H is also locally infinite.

In particular, t is always of infinite order in H. It is called the *stable letter* of the HNN extension.

The following result for HNN extensions is known as *Britton's Lemma* (Britton [1963]).

> Let $g_0, g_1, \ldots, g_n$ be a sequence of elements of G and let the letter ε, with or without subscripts, denote $+1$ or -1. A sequence $g_0, t^{\varepsilon_1}, g_1, t^{\varepsilon_2}, \ldots, t^{\varepsilon_n}, g_n$ will be called reduced if there is no consecutive subsequence t^{-1}, g_i, t with $g_i \in A$ or t, g_j, t^{-1} with $g_j \in B$. For a reduced sequence, and $n \geq 1$, the element
>
> $$g_0 t^{\varepsilon_1} g_1 t^{\varepsilon_2} \cdots t^{\varepsilon_n} g_n$$
>
> of H is different from the unit element.

The construction of an HNN extension can be generalized by introducing more than one stable letter and more than one isomorphism of subgroups, and Britton's Lemma can be put in a different form, resulting in a unique normal form for the elements of H. But these are minor improvements. To substantiate our remark about the applications of the HNN extension, we mention here the following results from the original HIGMAN, NEUMANN, and NEUMANN paper:

Every group G can be embedded in a group G^* in which all elements of the same order are conjugate. In particular, every torsion-free group can be embedded in a group G^{**} with only two conjugacy classes. If G is countable, so is G^{**}. (The only group containing elements of finite order with only two conjugacy classes is the group of order 2). Also, every countable group G can be embedded in a group H generated by only two elements. If the number of defining relations for G is n, the number of defining relations for H can be taken to be n. However, the paper by B. H. NEUMANN [1937b], mentioned above, shows that not all countable groups can be embedded in a single two-generator group since such a group would have only countably many two-generator subgroups, whereas the number of these is noncountable.

The construction of infinite groups with only two conjugacy classes illustrates the great power of the HNN extension. As far as embedding theorems are concerned, the main surprises came later. They are based on a fundamental theorem proved by G. HIGMAN [1961] which, at the same time, exhibits the fact that certain concepts and ideas developed in mathematical logic are of intrinsic importance for combinatorial group theory. A brief report on these results will be given in Chapter II.11.

We conclude this chapter by mentioning that both the free product with amalgamations and the HNN extensions admit fruitful topological interpretations. The free product with amalgamated subgroups can be interpreted as the fundamental group of a space which arises from spaces which have as fundamental groups the factors of the product by identifying certain subspaces. And an HNN extension of a group G can be interpreted as the fundamental group of a space which arises from a space with G as its fundamental group by attaching a one-dimensional handle. For the precise definition of these constructions, see LYNDON and SCHUPP [1976, pp. 179–188]. They are based on a topological theorem proved by VAN KAMPEN [1933a and 1933b] and usually named after him, although MASSEY [1967, p. 111] calls it the Seifert–van Kampen Theorem.

The latest use of graphs in combinatorial group theory which contains unified theories of the free products with amalgamations and of HNN extensions is the theory of groups acting on trees which has been presented in a monograph by J.-P. SERRE [1977].

Chapter II.5

One-Relator Groups

According to a theorem mentioned at the end of Chapter II.4, every finitely presented group can be embedded in a two-generator group with the same number of defining relations. This shows that, at least for the word problem, the number of generators of a group is immaterial if it is at least two. Intuitively, this fact must have been known almost from the beginning of the theory of groups given by presentations. It is the defining relations which make even the word problem so difficult. Indeed, for free (i.e., for no-relator) groups, the solution of the word problem was at least intuitively obvious already to DYCK in 1882. But for nearly half a century, the word problem for one-relator groups had been solved mainly with geometric methods for some knot groups and for fundamental groups of two-dimensional manifolds until SCHREIER [1927a] observed that his theorem on free products with amalgamations permits its solution for the class of one-relator groups mentioned in Chapter II.4. This paper contains practically all of the earlier examples.

However, it was not the word problem but a theorem about subgroups of one-relator groups which started their systematic investigation. It is due to DEHN, who called it the *Freiheitssatz*. The German word is now widely accepted as a technical term. A correct English equivalent would be "freeness theorem." It can be stated as follows.

Let a_ν ($\nu = 1, 2, \ldots$) be a set of generators of a group and let R be a freely and cyclically reduced word in the generators (which means that R does not contain any subsequences $a_\nu a_\nu^{-1}$ or $a_\nu^{-1} a_\nu$ and that the first and the last symbol in R form a pair distinct from the pairs a_ν, a_ν^{-1} and a_ν^{-1}, a_ν for any ν). Then in the group S with generators a_ν and defining relation $R = 1$, any subset of the generators which does not contain all of those a_ν which appear in R freely generates a free group.

This theorem is the strict analog of a theorem which says that an irreducible algebraic equation in which n complex variables actually appear cannot be used to derive any irreducible algebraic equation in which not all of these variables also appear.

As far as we know, there does not exist a proof (or even a sketch of a proof) written by DEHN for the *Freiheitssatz*. However, this statement has to be supplemented by the remark that many of the papers left by DEHN are very hard to decipher. This is true even for his handwritten papers using the ordinary alphabet. But many of the notes left by him are written in shorthand. DEHN's papers are now deposited in a Dehn Archive at the University of Texas, Office of the Librarian, Humanities Research Center, Box 7219, Austin, Texas 78712. However, we know something about DEHN's approach to the problem. Apparently, DEHN felt that his proof was not in a form suitable for publication, and in July 1928 he gave MAGNUS the proof of the *Freiheitssatz* as a topic for a Ph.D. thesis, sketching the method he had used as follows:

Suppose one generator, say t, of the group S actually appears in R but with exponent sum zero. Let $a, b, c, \ldots$ be the other generators of S and let S_0 be the subgroup which they generate. Then DEHN visualized the graph of S as a layered structure where each layer consists of the graph of one of the subgroups $t^n S_0 t^{-n} = S_n$ for $n \in \mathbb{Z}$. The union of these layers (which, of course, were not completely disjoint) would then form the graph of a normal subgroup N of S with the powers of t as coset representatives. The problem was to prove that each layer was a tree.

The proof of the *Freiheitssatz* given by MAGNUS [1930] is purely algebraic. Yet, there remains one element of DEHN's approach in his proof. It is the use of the normal subgroup N defined above. Suppose that the generator t actually appears and has exponent sum zero in R and that the *Freiheitssatz* has already been proved for all one-relator groups S^* with a relator R^* of length less than the length of R. Then the subgroup N has the following structure: It contains an infinite set of subgroups S_n^*, where n runs through the integers, each of which is isomorphic with the group S^*. N can be described as the infinite generalized free product of the S_n^*, where, for all n, free subgroups of S_n^* and S_{n+1}^* are amalgamated. SCHREIER's theorem then shows that in S the generators different from t freely generate a free subgroup. A rather sophisticated modification of this argument leads to the same result if t does not have exponent sum zero in R but another one of the generators appearing in R has this property. Finally, if all of the generators appearing in R have exponent sums $\neq 0$, one can reduce this case to the previous one by adjoining a suitable root of t to the group S and then carrying out a Nielsen transformation. Here "adjoining an mth root" means the following: We form the free product of S and an infinite cyclic group generated by an element τ and amalgamate t with τ^m. This has the effect that S is embedded in a larger group S' in which $t = \tau^m$ is now an mth power.

These remarks about the proof of the *Freiheitssatz* show the power of SCHREIER's theorem about free products with amalgamations. However, MAGNUS [1930] did not use SCHREIER's theorem but directly derived the special cases of it needed for his proof. Only in a footnote added when reading the galley proofs of his paper did MAGNUS observe that his lemmas were immediate consequences of SCHREIER's result.

Apart from the *Freiheitssatz*, DEHN had mentioned another problem to MAGNUS, and contributions to it also appear in MAGNUS [1930]. DEHN had called it the *root problem*, and in its simplest version it is the following question: Given a one-relator group S by a defining relator R, which words R' in the generators of S will have the property that $R' = 1$ implies $R = 1$? R' is then called a *root* of R. If, at the same time, R is also a root of R', then R and R' are called *equivalent*. MAGNUS proved that R' is then freely equal to a conjugate of $R^{\pm 1}$. In other words, if in a free group, $\{R\}$ denotes the normal closure of R, i.e., the normal subgroup generated by the conjugates of R, then, in turn, this normal subgroup determines the cyclically reduced pair of words $R^{\pm 1}$ uniquely. Examples of enumerations of all roots of some special relators are also given, and the problem of equivalence of pairs of relations is raised, in particular, the question whether equivalent pairs of relators can always be changed into each other by conjugations combined with Nielsen transformations. (These combinations were later called Q-*transformations* by RAPAPORT [1968].)

DEHN's root problem was taken up again much later, in a modified form, by A. STEINBERG [1969]. Independent of the theory of group presentations, R. BAER [1945] investigated the much more encompassing problem of representations of groups as quotient groups.

The adjunction of elements to groups which appears in MAGNUS [1930] in its simplest form and as a tool for the solution of another problem, has been introduced in its full generality by B.H. NEUMANN, [1943a]. He bases his proofs directly on SCHREIER's theorem. Of the many subsequent papers on this subject, we mention here only a few of the earlier ones. W. R. SCOTT [1951] used the concept of adjunction of elements for the introduction of the concept of an *algebraically closed group* and showed that every such group contains all finite groups. (In Chapter II.11 we shall encounter a remarkable improvement of this theorem.) And GERSTENHABER and ROTHAUS [1962] proved a sophisticated theorem about the possibility of adjoining solutions of r equations for r unknowns to a finitely generated group G which satisfies the condition of being isomorphic to a subgroup of a compact connected Lie group H. The group resulting from the adjunctions then also has an isomorphic replica in H. For the special case where the adjunctions consist only of adjunction of roots of elements, NEUMANN's interest in this topic also led to the Ph.D. thesis of G. BAUMSLAG [1960]. It contains a systematic theory of the following classes of groups: Let ω denote a nonempty set of prime numbers γ. The classes E, U, D are defined, respectively, as the classes of groups for which every element has, for every $\gamma \in \omega$, at least one γth root, at most one γth root, or precisely one γth root.

The use of the main result obtained by MAGNUS [1930], namely, the *Freiheitssatz*, started with two papers by MAGNUS [1931 and 1932]. The first one contains a solution of the word problem for a special class of one-relator groups, and the second one contains a solution of the word problem for all one-relator groups. It is an essential feature of this solution of the word problem and, more generally, of the solution of all word problems which utilize SCHREIER's theorem, that it requires the solution of what MAGNUS called an *erweitertes Identitätsproblem* and what is now known by the names of *occurrence* or *membership problem*. It was formulated by MIHAILOVA [1958] as follows: Let G be a group, and let W be an arbitrary word in its generators. Find an algorithm for deciding whether W represents an element of a finitely generated subgroup H whose generators are given explicitly as words in the generators of G. NIELSEN [1921] had solved this problem for free groups. In MAGNUS [1930 and 1932], it appears only in the special form where the generators of H form a proper subset of the set of generators of G. This special case is also known as the *generalized word problem*. Obviously, the solution of the word problem for the free product of two groups G_1 and G_2 with amalgamated isomorphic subgroups H_1 and H_2 requires precisely the solution of the membership problems for H_i in G_i ($i = 1, 2$).

A very brief sketch of the proof of the *Freiheitssatz* appeared in the book by REIDEMEISTER [1932b]. After that, the next and also the most important application of the theorem appears in a paper by LYNDON [1950] which was influenced by REIDEMEISTER who met LYNDON in Princeton in 1948. To explain one of LYNDON's results, we have to first describe the Ph.D. thesis of PEIFFER [1949] which had been instigated by REIDEMEISTER. PEIFFER starts with a free group Z with generators s_ν ($\nu = 1, \ldots, n$) and r_μ ($\mu = 1, \ldots, m$) and constructs the normal subgroup R^* which is generated by all of the conjugates

$$tr_\mu t^{-1},$$

where t denotes an arbitrary freely reduced word in the s_ν. The quotient group Z/R^* is then the free group S freely generated by the s_ν. Next, she defines a homomorphic mapping W of R^* onto a normal subgroup N of S. This is done by defining $W(r_\mu)$, for $\mu = 1, 2, \ldots, m$, as an arbitrary element $\neq 1$ of s and by postulating that

$$W\left(r_\mu^{-1}\right) = \left\{W(r_\mu)\right\}^{-1}, \qquad W\left(tr_\mu t^{-1}\right) = tW(r_\mu)t^{-1} \qquad \text{for all } t \in S$$

and, in general,

$$W(r_1^* r_2^*) = W(r_1^*) W(r_2^*) \qquad \text{for all } r_1^*, r_2^* \in R^*.$$

Clearly, S/N is then the group with generators s_ν and defining relations $W(r_\mu) = 1$, and N arises from the group R^* by replacing the r_μ by the words $W(r_\mu)$ in the s_ν.

The problem which now arises is the following one: to define N in terms of the generators of R^* as a quotient group R^*/K of R^* by a suitable set of defining relations for these generators. R^* is freely generated by $tr_\mu t^{-1}$, where $\mu = 1, \ldots, m$, and t runs through the complete set of freely reduced words in the s_ν. Let

$$\Omega\left(tr_\mu t^{-1}\right)$$

be a word in these generators. The mapping W maps Ω onto the element

$$\Omega\left(tW(r_\mu)t^{-1}\right)$$

of N. If Ω_k is a word of this type such that

$$\Omega_k\left(tW(r_\mu)t^{-1}\right) = 1,$$

then Ω_k is called a *relation identity*. All Ω_k are products of conjugates of the defining relators in a presentation of N in terms of the generators of R^*.

As a subset of these relators, we can always choose the particular Ω_k defined by

$$P = t_1^* r_\rho^{-1} \left(t_1^*\right)^{-1} t_2 r_\sigma t_2^{-1} t_1 r_\rho t_1^{-1} t_2 r_\sigma^{-1} t_2^{-1},$$

$$\rho, \sigma = 1, 2, \ldots, m,$$

where

$$t_1^* = W\left(t_2 r_\sigma t_2^{-1}\right) t_1$$

and where t_1, t_2 run independently through the freely reduced words in s_ν. The words P are called now *Peiffer identities*. The group R with generators s_ν, r_μ and defining relations $P = 1$ may or may not be the group N. One of the main results of her paper (PEIFFER [1949]) is the following theorem:

If $n > 1$, the group N is the quotient group of R with respect to its center.

There are other results which involve the effects of changing generators of R^* and also topological constructions which associate two-dimensional complexes with the various groups which she had introduced. We shall not describe these, and we have given a rather detailed account of the first part of her paper only because we could not find any report on it in the existing textbooks or monographs.

Returning now to the paper of LYNDON [1950], we can state his first main theorem as follows:

Consider a one-relator group with generators s_ν and a single defining relator $W(r_1) = 1$. If $W(r_1)$, as a freely reduced word in the s_ν, is not freely equal to a power V^e of another word V in s_ν, where $e > 1$, then the Peiffer identities define the complete set Ω_k of relation identities. If $e > 1$ and if e is maximal, then we have to add the relation identities

$$V^\varepsilon r_1 V^{-\varepsilon} r_1^{-1}, \qquad \varepsilon = 1, 2, \ldots, e - 1$$

to the Peiffer identities in order to obtain a complete system.

The second part of LYNDON's paper contains a computation of the cohomology groups of one-relator groups. We cannot give the details. A brief outline of the role of the cohomology theory of groups in combinatorial group theory may be found in Chapter II.9.

As an aside, we mention here that DEHN had described the problem of relation identities to MAGNUS in the early 1930's during a conversation. However, it has not been pursued either by DEHN or by any one of his students.

The next general theorem on one-relator groups was found by KARRAS, MAGNUS, and SOLITAR [1960]. Let S be a one-relator group with generators a_ν and a freely and cyclically reduced relator R. Assume that, in the free group generated by the a_ν, the word R can be written in the form

$$R = Q^e,$$

where the positive exponent e is maximal. Then, for $e = 1$, S has no elements $\neq 1$ of finite order. For $e > 1$, all elements $\neq 1$ of finite order are conjugates of powers of Q and the orders are divisors of e, and Q is exactly of order e. The methods used are the same as in MAGNUS [1932].

The last general theorem to be mentioned here is due to BAGHERZADEH [1975]:

Let M be a subgroup of S which is generated by a subset of the generators a_ν such that the Freiheitssatz applies to this subset. Let $s \in S$ be an element not contained in M. Then the intersection of M and $s^{-1}Ms$ is cyclic.

The theory of one-relator groups received a new impetus around 1960, in large part through the initiative of G. BAUMSLAG, who discovered that one-relator groups with torsion (i.e., with elements $\neq 1$ of finite order) can be investigated successfully with methods which do not apply to torsion-free groups. Also, G. BAUMSLAG, together with other authors, especially A. KARRASS and D. SOLITAR, uncovered the remarkable complexity of particular one-relator groups and systematically delineated the border between free groups and one-relator groups by showing which properties of free groups either are or are not shared by one-relator groups. One may look upon this development, which started thirty years after the proof of the *Freiheitssatz*, from two points of view. On the one hand, one may say that the *Freiheitssatz* and the subsequent solution of the word problem for one-relator groups by MAGNUS left the impression that one-relator groups are something almost as easy to deal with as free groups and that this fact delayed a more thorough investigation of one-relator groups for quite a while. This interpretation is due to G. BAUMSLAG and it certainly is at least a very plausible one. On the other hand, one may emphasize that DEHN, with uncanny intuition, had recognized exactly the one aspect of one-relator groups which they share fully with free groups.

We shall first describe our state of knowledge about one-relator groups with torsion. In order to avoid cumbersome repetitions, we shall (in this chapter only) denote these groups by the letter T. After that, we shall give some samples of one-relator groups with surprising properties.

In a paper whose topic had been suggested by G. BAUMSLAG, B. B. NEWMAN [1968] announced the following result which is now known as the *spelling theorem*:

Let T be the group with generators $b, c, d, \ldots$ and defining relator R, where R is freely and cyclically reduced and R is freely equal to Q^n, where $n > 1$. Let W be a freely reduced word which contains b and let V be a freely reduced word not containing b. Assume that W and V define the same element of T. Then W contains a subword which is identical to a subword of $Q^{\pm n}$ and of length greater than $(n - 1)q$, where q denotes the length of Q.

As a corollary, NEWMAN notes that the conjugacy problem for T is solvable. For torsion-free one-relator groups, this result has, until now (1980), been obtained only for rather restricted subclasses.

Another corollary of NEWMAN's result may be described as a remarkable generalization of the *Freiheitssatz* in the case of one-relator groups with torsion and only two generators b, c. In this case, the *Freiheitssatz* is trivial. But NEWMAN showed that unless R is a power of either b or of c, the elements b^γ and c freely generate a free subgroup of T whenever $\gamma > 2\beta$, where β is the largest absolute value of a b-exponent in R. A weaker version of this corollary has been derived by REE and MENDELSOHN [1974] using 2×2 matrix representations for T. MAGNUS [1975] showed that this method cannot yield the full result obtained by NEWMAN and that there are infinitely many groups T which cannot have a faithful matrix representation of this type. His proof uses the theorem of equivalent relators stated after the formulation of the *Freiheitssatz*.

NEWMAN's proof of his spelling theorem never appeared in print, but a brief proof due to McCOOL and SCHUPP appears on pp. 205–206 in LYNDON and SCHUPP [1977]. It would not make sense to reproduce here the collection of 31 theorems on general and special one-relator groups appearing on pp. 104–110 of the same monograph. But we shall cite two later papers by S. J. PRIDE [1977a and 1977b]. The first one of these describes the structure of all two-generator subgroups of any group T, and the second one contains a solution of the isomorphism problem for the class of two-generator groups T. This is by far the most comprehensive contribution to the difficult isomorphism problem which has been obtained so far (until 1980).

We conclude this chapter with some examples of one-relator groups which have unexpected properties. The first example is the group $S_{l,m}$ with generators a, b and defining relation

$$a^{-1}b^{l}ab^{-m} = 1.$$

BAUMSLAG and SOLITAR [1962] proved that whenever $l, m \geq 2$ are coprime, these groups are non-Hopfian, which means that they contain a normal subgroup $N \neq 1$ such that $S_{l,m}$ is isomorphic with $S_{l,m}/N$. It is trivial that there exist infinitely generated groups which are non-Hopfian. After HOPF [1930] raised the question whether there exist finitely generated and, in particular, finitely presented groups with this property, the Hopfian property of free finitely generated groups was established with algebraic methods by MAGNUS [1935] and was recognized as a corollary of results found by NIELSEN [1921], MAL'CEV [1940], and M. HALL [1950a]. But the first finitely generated non-Hopfian groups were constructed by B. H. NEUMANN [1950], and the first finitely presented such groups were discovered shortly afterwards by G. HIGMAN [1951]. It has three generators a, b, c and the two defining relations

$$a^{-1}ca = b^{-1}cb = c^{2}.$$

There still remained the question raised by B. H. NEUMANN, whether at least one-relator groups were Hopfian. The BAUMSLAG–SOLITAR examples show that this is not always true. In addition, the same paper has the result that the group with generators a, b and defining relation

$$a^{-1}b^{12}a = b^{18}$$

is Hopfian, although it contains the non-Hopfian group $S_{2,3}$ as a subgroup.

That the groups $S_{l,m}$ are far removed from free groups in other respects as well is shown by the remark in B. B. NEWMAN [1968] that the group $S_{2p,2}$ (where p denotes a prime number) contains an abelian subgroup which is isomorphic with the additive group of the p-adic rational numbers and, therefore, is infinitely generated. In a free group, the abelian subgroups are one-generator groups. (However, B. B. Newman also showed that a torsion-free one-relator group cannot contain the additive group of *all rational* numbers as a subgroup.)

At the other extreme, there exist one-relator groups which have so many properties in common with free groups that the fact that they are not free is rather surprising. We shall denote these groups by $S^*_{i,j}$, where i, j are nonzero integers. They are special cases of groups which G. Baumslag [1969] called *parafree* and they are defined as having three generators a, b, c and the single defining relation

$$a = c^{-i}a^{-1}c^i a c^{-j} b^{-1} c^j b.$$

The groups $S^*_{i,j}$ share the following properties with a free group F of rank 2.

(i) $S^*_{i,j}$ has a free normal subgroup with an infinite cyclic quotient group.
(ii) Every two-generator subgroup of $S^*_{i,j}$ is free.
(iii) Let $\gamma_n G$ denote the nth group of the lower central series of a group G. (For a definition, see Chapter II.7). Then the intersection of all of the $\gamma_n S^*_{i,j}$ is the unit element and

$$S^*_{i,j} / \gamma_n S^*_{i,j} \cong F/\gamma_n F \qquad \text{for all } n.$$

(iv) $S^*_{i,j}/S^{*\prime\prime}_{i,j} \cong F/F''$, where G'' denotes the second derived group of G.

The great difficulties of the isomorphism problem for torsion-free one-relator groups are well illustrated by the fact that until now (1980) there exists no proof showing that any two of the groups $S^*_{i,j}$ are nonisomorphic.

Probably the most surprising one of these properties is (iii). One way of explaining this is the following: Magnus [1939c] had proved that a group G, with $m + r$ generators and r defining relations, is the free group F_m of rank m if G can also be presented as a group on m generators. The proof is based on the fact that then $\gamma_n F_m / \gamma_{n+1} F_m$ and $\gamma_n G / \gamma_{n+1} G$ must be isomorphic for all n. One might therefore conjecture that, conversely, this sequence of isomorphisms suffices to prove that G is isomorphic with F_m if the intersection of the $\gamma_n G$ is the identity. Of course, the groups $S^*_{i,j}$ are counterexamples, disproving this conjecture.

Our knowledge of one-relator groups is by now broad enough to justify a monograph dealing exclusively with this topic. The existing textbooks and especially the classification of the summaries in the *Mathematical Reviews* (Volumes 1–40) by G. Baumslag [1974] clearly show how much information is now available. What we have tried to show is merely that the class of one-relator groups, although large enough to be interesting, is still sufficiently restricted to be an object of systematic investigation. There exist at present (1980) no indications whatsoever that a systematic theory of two-relator groups would be possible.

Chapter II.6

Metabelian Groups and Related Topics

The title of this chapter does not reflect its main purpose. We shall discuss metabelian groups here and not under the heading of "special groups" in Chapter II.10, where they would seem to belong, because their first investigation was motivated by problems arising in three different fields outside of group theory. Within a few years, FURTWÄNGLER [1930], REIDEMEISTER [1932a], and MOUFANG [1937] used metabelian groups to prove theorems pertaining to algebraic number theory, knot theory, and the foundations of geometry, respectively. We shall give some background information about these problems, starting with a few remarks about the authors, and, after discussing the particular theorems on metabelian groups involved in the solution of these problems, we shall give a brief account of some of the later developments arising from or related to the original investigations.

PHILIPP FURTWÄNGLER (1869–1940) studied in Goettingen during HILBERT's number-theoretical period. He made fundamental contributions to class field theory, a branch of algebraic number theory created by HILBERT. The "principal ideal theorem", the proof of which is the purpose of FURTWÄNGLER's paper to be discussed here, was the last of the still unproved general conjectures formulated by HILBERT. In his number-theoretical papers, HILBERT used group theory systematically, but all of the groups appearing in his papers are abelian. It is remarkable that FURTWÄNGLER had developed an active interest in nonabelian groups well before their applicability to number theory became known through the paper by ARTIN [1927]. To support this statement, we observe that the topic of the Ph.D. thesis of SCHREIER had been suggested to him by FURTWÄNGLER before 1923, and it deals with the general theory of group extensions. Incidentally, the distribution of the papers by FURTWÄNGLER over the many years of his mathematical productivity exhibits a phenomenon which is incompatible with the widely held view according to which mathematicians conceive their original ideas when they are young and, if they remain productive at all, use their later years to build on the foundations they had laid in their youth. But FURTWÄNGLER was 61 years old when his paper on the principal ideal theorem appeared, and the ideas and techniques on which his proof is based were developed only in the decade preceding his paper, partly through his initiative. Also, the theorem is far from being trivial. Otherwise, ARTIN, who discovered the group-theoretical formulation of the theorem and who was one of the most brilliant young mathematicians of his time, would not have failed to give a proof himself. Incidentally, ARTIN's name will appear again below in connection with this problem. An appraisal of the work of REIDEMEISTER has already been given in Chapter II.3.

RUTH MOUFANG (1905–1977) was a student of MAX DEHN. A large part of her work is dedicated to problems in the foundations of geometry. Her most outstanding contribution to this field is a result which adds a third important discovery to two others made previously by HILBERT [1901 and 1930]. Reversing a development going from EUCLID to DESCARTES in which geometry is replaced by algebra as the fundamental discipline of mathematics, HILBERT had shown that a subset of his axioms for plane geometry (essentially, the incidence axioms) together with the incidence theorem of DESARGUES permits the introduction of coordinates on a straight line which are elements of a skew field. If DESARGUES' theorem is replaced by that of PAPPUS, the coordinates become elements of a field. MOUFANG [1933] showed that another incidence theorem, called the theorem of the complete quadrilateral (or of the invariance of the fourth harmonic point), allows one to introduce coordinates which are elements of an alternating division algebra. This and a subsequent paper (MOUFANG [1934]) had the effect of stimulating further research of these algebras and of other nonassociative algebraic structures ("Moufang loops"). Her work is based both on a powerful geometric intuition and on the development of difficult algebraic techniques. It is supplemented by a sequence of papers in continuum mechanics. These were due to the fact that the German minister of education under the government of HITLER refused her application of admission to an academic career. She then joined an industrial firm, being one of the first mathematicians in Germany with a Ph.D. degree to do so and certainly the first woman in Germany to hold such a position. In 1946, she returned to academic life and became the first woman in Germany to be appointed as a full professor of mathematics. Her only paper in group theory will be discussed below, with an explanation of its motivation which arises from the foundations of geometry.

A. The Principal Ideal Theorem

As background information to the paper by FURTWÄNGLER [1930], we mention EDWARDS [1977], who gives a historical account of algebraic number theory up to the time of HILBERT's work and the appendix by H. HASSE in HILBERT [1932], which summarizes later developments up to 1930 and ends with a statement of FURTWÄNGLER's result. Here, we can give only a few definitions and a brief explanation of the reasons for which the principal ideal theorem has attracted so much interest.

Let K be a finite algebraic extension field of the field Q of rational numbers and let $R(K)$ be the ring of algebraic integers in K. In general, the elements of $R(K)$ cannot be factorized uniquely (apart from factors which are units) into a product of prime numbers. However, the ideals in $R(K)$ can be factorized uniquely into a product of prime ideals. The ideals in $R(K)$ are distributed over finitely many disjoint ideal classes which are defined as follows: Let j_1 and j_2 be two ideals in $R(K)$. Then j_1 and j_2 belong to the same ideal class if and only if there exist elements α and β of $R(K)$ such that

$$\{\alpha\}j_1 = \{\beta\}j_2.$$

Here $\{\alpha\},\{\beta\}$ denote the principal ideals generated by α and β, respectively (i.e., the ideals consisting, respectively, of the products of α and β with arbitrary elements of $R(K)$).

Consider now a field K^* which is a finite normal (or Galois) extension of K. Let n be the degree of K^* with respect to K. A prime ideal π in $R(K)$ will then either factor into d distinct prime ideals in $R(K^*)$, where d is a divisor of n, or π will be divisible by a power π^{*d} with $d > 1$ of a prime ideal π^* in $R(K^*)$. In the latter case we say that K^* is *ramified* over K. (This notation arises from the algebraic theory of Riemann surfaces, i.e., from the theory of finite algebraic extensions E of the field of rational functions of a complex variable with coefficients in the field $\mathbb{C}$ of complex numbers. A point on the Riemann surface can then be described as an ideal in a subring contained in E.)

Class field theory deals with the case where the Galois group of K^* over K is abelian and it describes the factorization of the ideals π of $R(K)$ in $R(K^*)$. The most elementary example for this beautiful but complex theory can be formulated as Legendre's theorem according to which an odd prime number p is the sum of two squares of natural numbers if and only if $p \equiv 1 \bmod 4$. (In this case, $R(K)$ is the ring $\mathbb{Z}$ of natural integers and $R(K^*)$ is the ring $\mathbb{Z}(i)$ of Gaussian integers.) Here we cannot try to give anything resembling a full description of class field theory, but now we can at least formulate the principal ideal theorem. It states:

Given K, there exists a maximal unramified Galois extension K^* of K with an abelian Galois group for K^*/K. In $R(K^*)$, every ideal of $R(K)$ becomes a principal ideal.

The simplest example for this theorem is provided by the case where K arises from Q by adjoining $\sqrt{-5}$. The ring $R(K)$ then has two ideal classes. By adjoining $\sqrt{-1}$ to K, we obtain an unramified extension K^* such that in $R(K^*)$ every ideal generated by elements of $R(K)$ becomes a principal ideal. (As it happens, all ideals in $R(K^*)$ are principal ideals according to HILBERT [1932, p. 51].)

The principal ideal theorem shows that the necessity of dealing with ideals rather than numbers in order to obtain a unique factorization theorem may be bypassed by embedding the ring $R(K)$ into a larger ring. (Of course, the computational advantages connected with this discovery are minimal since practically all rings $R(K)$ contain infinitely many units, the only exceptions being the integers and the cases where K is an imaginary quadratic field.) Now it is not true in general that the number of ideal classes in the larger ring is actually equal to 1, and in this respect the example given above is misleading. There arises the question raised by FURTWÄNGLER and known as the *class field tower problem*: whether a finite number of repetitions of the process described above which leads from K to K^* will eventually lead to a ring in which all ideals and not only those generated by elements of $R(K)$ will become principal ideals. But, in general, this will not happen. In a beautiful paper, GOLOD and ŠAFAREVIČ [1964] gave a criterion for an algebraic number field K which, if satisfied, establishes the fact that K cannot be embedded in any other algebraic number field K^{**} of finite degree over Q such that in $R(K^{**})$ every ideal is a principal ideal. GOLOD and ŠAVAREVIČ also show that this is already true for certain imaginary quadratic number fields, i.e., for fields which arise from Q by adjoining $\sqrt{-D}$, where D is a positive integer. As an example, they prove this for the case where

$$D = 3 \cdot 5 \cdot 7 \cdot 11 \cdot 13 \cdot 17 \cdot 19 = 4849845.$$

KOCH [1969] has shown that there already exist infinitely many such fields where D is a product of only four prime numbers.

The paper by GOLOD and ŠAFAREVIČ [1964] contains several algebraic theorems which are needed to prove their number-theoretical result. Some of these are purely group theoretical and will be formulated after a description of FURTWÄNGLER's paper to which we now turn.

FURTWÄNGLER [1930] starts with an acknowledgement of the papers by ARTIN [1927] and SCHREIER [1926a and 1926b] which contain, respectively, the group-theoretical formulation of the principal ideal theorem and the basic theory of group extensions. He then defines a *metabelian group M* as a group with an abelian commutator subgroup M'. This term is now standard, but in the older literature the term has been used by some authors to denote groups in which the quotient group of the center is abelian. These groups would now be called *nilpotent of class* 2. They form a true subclass of the class of metabelian groups.

FURTWÄNGLER uses the following notation which has been standard ever since and which, according to him, is suggested by a notation customary in number theory: Let a, b be any two elements of any group G, then $b^{-1}ab$ will be denoted by a^b. Only finitely generated groups M for which M/M' is finite will be needed. The abelian group $A = M/M'$ will then be a direct product of finitely many finite cyclic groups.

Let s_i, $i = 1,\ldots,n$ be a set of elements of M such that the products

$$s = s_1^{m_1} s_2^{m_2} \cdots s_n^{m_n}, \qquad 0 \leqslant m_i < e_i, \qquad i = 1,\ldots,n$$

form a complete set of distinct coset representatives of M' in M, where we also assume that

$$s_i^{e_i} = t_i \in M'.$$

For any element $t \in M'$, t^s depends only on the coset sM'. We can now introduce the group ring $\mathbb{Z}(A)$ of the abelian group A over the integers by mapping

$$s_i^{m_i} M' \to \sigma_i^{m_i}, \qquad sM' \to \sigma_1^{m_1} \sigma_2^{m_2} \cdots \sigma_n^{m_n}, \qquad M' \to 1,$$

where the σ_i are now generators of a commutative ring in which

$$\sigma_i^{e_i} - 1 = 0.$$

For any element $\alpha \in \mathbb{Z}(A)$, we can define t^α, where α is now a polynomial in $\sigma_1,\ldots,\sigma_n$ with integral coefficients and of degree $\leqslant e_i - 1$ in σ_i. And t^α is uniquely defined because for any $\alpha, \beta \in \mathbb{Z}(A)$ and any $t \in M'$, we have

$$t^\alpha t^\beta = t^\beta t^\alpha, \qquad (t^\alpha)^\beta = (t^\beta)^\alpha.$$

Therefore, we can consistently define

$$t^{\alpha+\beta} = t^\alpha t^\beta = t^\beta t^\alpha = t^{\beta+\alpha},$$

$$t^{m\alpha} = (t^m)^\alpha = (t^\alpha)^m,$$

$$t^{\alpha\beta} = (t^\alpha)^\beta = t^{\beta\alpha}.$$

Now we are finally able to formulate FURTWÄNGLER's result which implies the principal ideal theorem for algebraic number fields. It states:

Let

$$f_i = 1 + \sigma_i + \sigma_i^2 + \cdots + \sigma_i^{e_i - 1}$$

and assume that M is generated by the s_i, $i = 1, 2, \ldots, n$. Then, for any choice of the $t_i = s_i^{e_i} \in M'$,

$$t_i^{f_1 \cdots f_{i-1} f_{i+1} \cdots f_n} = 1.$$

Here, of course, it should be noted that t_i will satisfy the relation

$$t_i^{\sigma_i - 1} = t_i^{s_i} t_i^{-1} = 1.$$

FURTWÄNGLER's proof consists of nearly 20 printed pages of ingenious calculations. We shall mention other proofs later on, but first we shall sketch the history of the *group tower problem* which arose from the class field tower problem, and we shall describe the group-theoretical result found and used by GOLOD and ŠAFAREVIČ in their construction of infinite class field towers.

In its simplest version, the group tower problem can be formulated as follows: Let P be a finite p-group, that is, a group in which the order of every element is a power of a fixed prime number p. In such a group, the sequence of derived groups, $P, P', P'', \ldots$ in which each term is the commutator subgroup of the previous group necessarily terminates after a finite number n of steps with the unit element, i.e., $P^{(n)} = 1$ for sufficiently large n. The question is: Given the abelian group $A = P/P'$, can it be that there exists an upper bound for the number n which depends only on the structure and order of the group A? This is indeed true if A is cyclic, in which case $n = 1$, or if A is of order 4, in which case $n \leqslant 2$. But according to SERRE [1964], it is not true in any other case. This paper by SERRE terminates a sequence of investigations of the problem, some of which are partly number theoretical (SCHOLZ and TAUSSKY [1934]) and most of which are purely group theoretical (TAUSSKY [1937], MAGNUS [1935], ITO [1950], to mention only the earlier ones).

In spite of these group-theoretical results, it still could be true that the class field tower must always terminate after a finite number of steps. That this is not true either follows from two theorems, one of which is purely group theoretical and is due to GOLOD and ŠAFAREVIČ [1964]. It states:

Let P be a finite p-group. Let $d(P)$ be the minimal number of generators of P (which, according to BURNSIDE [1913], always equals the minimal number of generators of $A = P/P'$) and let $r(P)$ denote the minimal number of defining relations needed for a presentation of P. Then

$$r(P) > \tfrac{1}{4}(d(P) - 1)^2.$$

To obtain now a construction for an algebraic number field K for which the class field tower does not terminate, a theorem of ŠAFAREVIČ [1963] is needed. Considering only the p-component (which we shall not define here) of the group of the class field tower, we obtain either a finite p-group P if the tower terminates or a so-called *pro-p-group* P^*.

This is the inverse limit (as defined in Chapter II.9) of an infinite sequence of finite p-groups. Its minimal number d of generators is again that of the p-component of the (abelian) group of the class field K^* with respect to K. It is possible to define a finite number $r^*(P^*)$ for P^* as the analog of the minimal number $r(P)$ of defining relations for a finite p-group P. If we have a finite class field tower, $r = r^*$.

Finally, let ρ be the minimal number of generators of the multiplicative group of units in K. These are the algebraic integers ϵ in K for which ϵ^{-1} also is an algebraic integer. Then ŠAFAREVIČ [1963] proved

$$r^* \leqslant d + \rho$$

and observed that the inequality stated above in the theorem of GOLOD and ŠAFAREVIČ implies that the group of the class field tower must be infinite if

$$\tfrac{1}{4}(d-1)^2 \geqslant d + \rho.$$

An explicit construction of imaginary quadratic number fields with infinite class field tower can now be obtained from number-theoretical results going back to GAUSS.

We shall not mention here various papers connected with (and, in part, improvements of) the results obtained by GOLOD and ŠAFARAVIČ. But we wish to comment on the importance of these results. The use of group theory (including nonabelian groups) in the theory of algebraic number fields is, of course, nothing new. CAMILLE JORDAN characterized the work of GALOIS by saying that "group theory is the metaphysics of the theory of equations," and HILBERT's *Zahlbericht* contains numerous uses of group theory in algebraic number theory. But the theorem of GOLOD and ŠAFAREVIČ is the first important theorem valid for a large class of groups in which the minimal number of defining relations for presentations of the groups enters as an essential concept. And the construction of a pro-p-group is based on the introduction of topological concepts into the theory of countable groups which became possible only after the older topological concepts of "open set" and "neighborhood" arising from point-set topology had gone through a process of distillation by abstraction. We believe that this cooperation of ideas from different fields makes the theorem of GOLOD and ŠAFAREVIČ particularly noteworthy.

Now we shall return to the proof of the principal ideal theorem. Commonly, first proofs are difficult and can be simplified later. In 1932, SIEGEL suggested to MAGNUS that simplifying FURTWÄNGLER's proof would be a worthwhile task. As it happened, MAGNUS at this time had been interested in another question concerning metabelian groups which had been raised by the topologist H. KNESER, who had asked whether there exist two-generator groups of euclidean motions of the plane which are not finitely related. Any such group is metabelian, and it can be represented by a pair of matrices of the form

$$\begin{pmatrix} g, & t \\ 0, & 1 \end{pmatrix},$$

where g, t are complex numbers and $|g| = 1$. Using the methods of MAGNUS [1934b], it is not difficult to show that even the free metabelian groups with finitely many generators (i.e., the groups with defining relations which express the fact that any two commutators commute) can be represented faithfully in terms of such matrices. But to show that such a group must be infinitely related is surprisingly difficult. It was proved, within a broader context, by SMELKIN [1965a]. However, MAGNUS [1934b] succeeded in simplifying the proof of the principal ideal theorem by showing that the metabelian group M introduced earlier in this chapter has a faithful matrix representation defined by the mapping

$$s_i \rightarrow \begin{pmatrix} \sigma_i, & t_i \\ 0, & 1 \end{pmatrix},$$

where t_i are indeterminates. The general element s of M is then represented by a matrix

$$\begin{pmatrix} \sigma, & L \\ 0, & 1 \end{pmatrix},$$

where L is a linear form in the t_i with coefficients from the group ring of the abelian group $A = M/M'$ (generated by σ_i) and σ is a product of σ_i, i.e., an element of A. The proof is based on the Reidemeister–Schreier theorem. It can easily be generalized to fit more complicated situations. In its most general form, it reads as follows: Let G be a group with generators g_ν, where ν runs through any set of subscripts. Let F be the free group with free generators g_ν^*. Then G is a quotient group of F under a homomorphic mapping defined by $g_\nu^* \rightarrow g_\nu$. Let R be the kernel of this mapping and let R' be the commutator subgroup of R. Then the group F/R' has a faithful representation as a group of 2×2 matrices with zeros in the left lower corner and is generated by the particular matrices

$$\begin{pmatrix} g_\nu, & t_\nu \\ 0 & 1 \end{pmatrix},$$

where the t_ν are indeterminates commuting with all g_ν. The general element of F/R' is then of the form

$$\begin{pmatrix} g, & L_g \\ 0 & 1 \end{pmatrix},$$

where g is an element of G and L_g is a linear form in (finitely many) t_ν with coefficients from the group ring $\mathbb{Z}(G)$ of G with integral coefficients. This theorem was proved for finite groups G by MAGNUS [1939b], who used it to simplify the proof of a theorem by M. HALL [1938] in the theory of group extensions, and was proved in full detail for arbitrary groups G by M. HALL [1959, p. 23]. The effectiveness of this result is based on the fact that the abelian group R/R' (which is a $\mathbb{Z}(G)$-module) is now embedded in a *free* $\mathbb{Z}(G)$ module with basis elements t_ν. Not all elements of this free module appear as forms L_g, but calculations become easier now and more transparent, and consequently the proof of FURTWÄNGLER's theorem could be reduced to four pages by MAGNUS [1934b]. There exist various later utilizations of this method, e.g., by BACHMUTH [1965, 1966], and there exist generalizations involving $n \times n$ $(n > 2)$ rather than 2×2 upper triangular matrices. The latest one of these, which also contains references to earlier work, is the paper by K. G. GUPTA and N. D. GUPTA [1978].

We note here that the matrix representation for F/R' described above is connected with an important technique which was developed later and the rudiments of which we shall now sketch.

In 1953, R. FOX started a sequence of papers with the title *Free differential calculus*, introducing a new method into combinatorial group theory which has been widely used since. His basic definitions are the following:

Let F be a free group on free generators x_i, $i = 1, 2, \ldots$, and let $\mathbb{Z}F$ denote the group ring of F with integral coefficients. Let ϵ be the mapping of $\mathbb{Z}F$ onto the ring $\mathbb{Z}$ of integers, defined by $x_i \to 1$, where 1 now denotes both the integer 1 and the unit element of F. A *derivation* D is defined as a mapping of $\mathbb{Z}F$ into $\mathbb{Z}F$ with the following properties: Let u, v be any words in the x_i. Then

$$D(uv) = Du + uDv, \qquad D1 = 0;$$

$$D(mu + nv) = mDu + nDv, \qquad m, n \in \mathbb{Z}.$$

A derivation D is uniquely determined by the Dx_i which can be chosen arbitrarily. We can express Du explicitly in terms of the Dx_i by introducing the *Fox derivatives* $\partial u / \partial x_i$ which are defined as follows: First consider the case where $u = 1$ or $u = x_i^{\pm 1}$, and put

$$\frac{\partial 1}{\partial x_i} = 0, \qquad \frac{\partial x_i}{\partial x_i} = 1, \qquad \frac{\partial x_i^{-1}}{\partial x_i} = -x_i^{-1}, \qquad \frac{\partial x_j^{\pm 1}}{\partial x_i} = 0, \text{ if } i \neq j.$$

Next, assume that u is a product

$$u = y_1 y_2 \cdots y_n$$

of n factors taken from the sets of the x_i and the x_i^{-1}, and define

$$\frac{\partial u}{\partial x_i} = \sum_{\nu=1}^{n} y_1 \cdots y_{\nu-1} \frac{\partial y_\nu}{\partial x_i}.$$

Then, for any derivation D,

$$Du = \sum_i \frac{\partial u}{\partial x_i} Dx_i.$$

A special derivation is given by the mapping $u \to u - \epsilon(u)$. It is given by

$$u - \epsilon(u) = \sum_i \frac{\partial u}{\partial x_i}(x_i - 1)$$

and Fox called the left ideal generated by the $x_i - 1$ the *augmentation ideal* of $\mathbb{Z}F$.

The connection of this construction with the matrix representation F/R' described above is established by the following formulas:

Let u be a word in x_i. Replace x_i and x_i^{-1}, respectively, in u by the matrices

$$M_i = \begin{pmatrix} x_i, & t_i \\ 0, & 1 \end{pmatrix}, \qquad M_i^{-1} = \begin{pmatrix} x_i^{-1}, & -x_i^{-1}t_i \\ 0, & 1 \end{pmatrix},$$

where t_i are indeterminates forming a basis for a free module over the ring $\mathbb{Z}F$. Now replace $x_i^{\pm 1}$ in u by $M_i^{\pm 1}$. The result is a matrix

$$\begin{pmatrix} u, & \sum_i \partial u/\partial x_i t_i \\ 0, & 1 \end{pmatrix}.$$

In other words, the Fox derivatives appear as coefficients of the t_i.

It is not clear when this connection was observed for the first time. It is mentioned in print by LYNDON and SCHUPP [1977, pp. 70–71], but it seems that for many years it was not noticed by anyone (including FOX and MAGNUS).

The paper by MAGNUS [1934b] was followed immediately by a paper by IYANAGA [1934], which contains a new proof of the principal ideal theorem. The importance of this proof is due not so much to its greater brevity (it is just a little shorter than that of MAGNUS), but to the fact that it is based on the use of concepts and theorems which are of interest in a broader context. (IYANAGA himself commends his proof because it makes no use of the Reidemeister–Schreier method. In 1934, this was a correct assessment. Today (1980), the method is considered elementary.)

IYANAGA starts with a reformulation of the principal ideal theorem as a theorem about a *transfer* (*Verlagerung*). Avoiding technicalities as far as possible, this concept can be introduced as follows (see ZASSENHAUS [1956, pp. 164–168]): Let G be a group with a subgroup H of finite index n. We can construct a homomorphic image (called transfer) of G as a subgroup of the abelian group H/H' as follows: Let r_i, $i = 1, \ldots, n$ be right coset representatives of H in G and let g be an element of G. Then

$$r_i g = h(i, g) r_{j(i)},$$

where $j(i)$ defines a permutation of the symbols $i = 1, \ldots, n$ and where $h(i, g)$ is an element of H depending on g. If the $h(i, g)$ would be real numbers, this would be considered as a *monomial matrix representation* of G, i.e., a matrix representation which arises from a representation by permutation matrices if we replace the nonzero elements (which equal 1) in the permutation matrix by real numbers. The absolute value of the product of these real numbers would then be the absolute value v of the determinant of the matrix, and mapping g onto v would then produce a one-dimensional representation of G with entries from the positive real numbers. As it is, the $h(i, g)$ are elements of a group which need not even be commutative. However, if we define $\pi(g)$ by

$$\pi(g) = \prod_{i=1}^{n} h(i, g) \bmod H'$$

as an element of H/H', it is uniquely determined, and the same argument which shows that the absolute values of the determinants of monomial representations over the real numbers deliver a one-dimensional representation of G now also proves that the mapping τ defined as

$$\tau: g \to \pi(g)$$

is a homomorphic mapping of G into H/H'. This mapping τ is called the *transfer* of G into H, and the principal ideal theorem can now be stated as follows.

Let M be a metabelian group generated by the coset representatives of M', and assume that M/M' is finite. Then the transfer of M into M' is of order 1.

The concept of transfer is of great importance in the theory of finite groups. As the definitions above show, it is not strictly confined to this theory, but it plays hardly any role in the theory of infinite groups since its definition requires the index of H in G to be finite. For this reason, it will not appear again in our historical account.

The next concept appearing in IYANAGA's paper is that of the *order-ideal* (*Ordnungsideal*). If we write the group M' additively, it can be considered as a $\mathbb{Z}(A)$ module where, again, $A = M/M'$ and where $\mathbb{Z}(A)$ is the group ring of A over $\mathbb{Z}$ (the integers). Let μ_ρ, $\rho = 1,\ldots,r$, be a minimal system of elements in M' such that every element in M' can be expressed as a sum

$$\sum_{\rho=1}^{r} \alpha_\rho \mu_\rho, \qquad \alpha_\rho \in \mathbb{Z}(A)$$

and form all $r \times r$ determinants of systems of r linear relations of the form

$$\sum_{\rho=1}^{r} \alpha_\rho^* \mu_\rho = 0.$$

Then in $\mathbb{Z}(A)$ these determinants generate an ideal Ω and every element ω of Ω has the property that $\omega\mu = 0$ for all elements μ of M'. This ideal turns out to be independent of the choice of μ_ρ and is called the order-ideal of M'. We cannot describe IYANAGA's method of computing Ω and mention only that he proved the principal ideal theorem by showing that the products

$$f_1 \cdots f_{i-1} f_{i+1} \cdots f_n$$

(which we introduced in FURTWÄNGLER's formulation of the theorem) are elements of Ω.

The word "order-ideal" is one of the many terms which have been introduced into the language of algebra during the Twentieth Century. They make it possible to be very concise when formulating results. In our definition of the order-ideal we have used another such term, namely, the word $\mathbb{Z}(A)$-module. It does not appear in the paper of IYANAGA, who gives an explicit description of it.

The third important concept introduced by IYANAGA [1934] is that of a *splitting extension* of a group. It is due to ARTIN, who also had, in lecture notes, given a proof of a basic theorem about the existence of such an extension which appears for the first time in print in IYANAGA's paper. We shall report on splitting extensions in Chapter II.9.

The range of validity of the principal ideal theorem has been analyzed by WITT [1936]. A different proof has been given by SCHUMANN [1938]. He utilizes topological ideas arising from the concept of the graph of a group which had been developed by Schumann [1937] and analyzes the connection between his proof and that given by IYANAGA. Also, in Section 4 of his paper, he constructs for any finite group G with a presentation $G = F/R$ (F free) a representation for F/R' in terms of pairs of symbols for which a binary associative composition is defined. It can be shown that these pairs of symbols can be identified with the elements in the first rows of the matrix representations given by MAGNUS [1934b and 1939b].

B. Applications to the Theory of Knots and Links

The interest of SCHUMANN in metabelian groups as mentioned above certainly was not incidental. A few years earlier, he and REIDEMEISTER had used metabelian groups to obtain "invariants" of knots and links. These are algebraic quantities which are computable from the projection of one or several sufficiently smooth nonintersecting curves in three-space onto a plane and do not change if we replace the curve or curves by an isotopic system.

In 1928, the topologist J. W. ALEXANDER had found invariants for knots and links by using purely geometric methods. In the case of a knot, the invariant he found is what he called an *L-polynomial*. The "*L*" stands for "Laurent", and an *L*-polynomial is a finite laurent series in an indeterminate x with integral coefficients l_n, i.e., an expression

$$L(x) = \sum_{n=-N}^{+N} l_n x^n,$$

where N is arbitrary but finite. REIDEMEISTER [1932a] showed that the *L*-polynomial of a knot can in fact be computed from a presentation of the fundamental group K. Such a group is finitely presented and the abelian group $C = K/K'$ is always infinite cyclic. Therefore, although isotopic knots have isomorphic fundamental groups (a fact mentioned by ALEXANDER [1928], who adds that it appears to be very difficult to utilize), it is not possible to distinguish between any two of them by just abelianizing the groups. However, one may try to use other quotient groups of K which must be isomorphic for isomorphic groups, and one such possibility is given by the quotient group of the second commutator subgroup K''. Then K/K'' is the maximal metabelian quotient group of K. Its commutator subgroup is the abelian group $A = K'/K''$. In general, A will be infinitely generated. Nevertheless, it has a finite structure, due to the fact that K is finitely generated. If we write A additively, and if we introduce the group ring $\mathbb{Z}(C)$ of the infinite cyclic group C over the integers $\mathbb{Z}$, then there exist finitely many elements α_λ in A such that every element of A can be written as a sum

$$\sum_\lambda \gamma_\lambda \alpha_\lambda, \qquad \gamma_\lambda \in \mathbb{Z}(C).$$

In other words, A is a finitely generated $\mathbb{Z}(C)$-module. Now $\mathbb{Z}(C)$ is exactly the ring of all *L*-polynomials in the indeterminate x if we define C as the multiplicative group of the powers x^n of x. Those elements γ of $\mathbb{Z}(C)$ for which $\gamma \alpha_\lambda = 0$ for all λ form an ideal J in $\mathbb{Z}(C)$ (which actually is the "order-ideal" as defined by IYANAGA [1934]). Since $\mathbb{Z}(C)$ is a principal ideal ring, all elements of J are multiples of a single element L_K which, however, is not uniquely determined since we can replace it by any element $\pm x^n L_K$. (The elements $\pm x^n$ are the "units" of $\mathbb{Z}(C)$, i.e., the elements which have an inverse within $\mathbb{Z}(C)$). One usually normalizes L_K so that the coefficient of the highest power of x occurring in L_K is positive and that the term $\neq 0$ of lowest degree in L_K is the one of degree zero. This normalized L_K is then called *the L-polynomial* (and also the *Alexander polynomial*) of K.

In his book, REIDEMEISTER [1932a] pays homage to ALEXANDER by demonstrating the invariance of L_K as a property of a knot (independent of its projection onto a plane) both geometrically and algebraically. Undoubtedly, the algebraic derivation has considerable methodological merit, and at the time REIDEMEISTER found it, the algebraic methods he used were not yet standard ones. In fact, the use of characteristic quotient groups other than that of the commutator subgroup for the investigation of groups given by a presentation is largely due to REIDEMEISTER and may be considered as one of his most important achievements. We shall mention other contributions by him of the same type in the next chapter. As for ALEXANDER's paper, the discovery of the L-polynomial as a knot invariant is an astonishing feat. How does one get the idea of associating a polynomial with the projection of a knotted curve on a plane?

ALEXANDER [1928] also found L-polynomials in two indeterminates which are invariants of pairs of linked curves in three space both of which may themselves be knots. REIDEMEISTER and SCHUMANN [1934] derived these invariants from the structure of the maximal metabelian quotient group of the fundamental group of the linkage. The method is not confined to linkages of two curves, but the results are much more technical than in the case of a single curve. We shall not describe them here and we shall even omit references to later papers dealing with this topic. (For the years 1940–1970, Chapter 23 in BAUMSLAG [1974] provides a good survey of the literature.) However, we shall now report briefly on a paper by SEIFERT which also appeared in 1934. It has nothing to do with metabelian groups except for the fact that it introduces a new method for the investigation of knot groups which may still work in those cases where the computation of their metabelian quotient groups provides no information at all.

If the group K of a knot is infinite cyclic, we have $L_K = 1$. This will happen if the knot is isotopic with a circle, but there are also cases where $L_K = 1$, although K is the fundamental group of a truly knotted curve and, according to Dehn's lemma, not cyclic. To prove this, one cannot use any solvable quotient groups of K since, automatically, all of these would be cyclic again. Now SEIFERT [1934] had constructed the first example of a knot with the Alexander polynomial $L_K = 1$. In order to prove that its group was not infinite cyclic, he mapped it homomorphically onto an infinite group of 2×2 matrices with determinant ± 1. Actually, he did not even have to compute the matrices since he could show that his knot group had a quotient group with a presentation known to be that of an infinite fuchsian group. One may prefer to call this argument a clever trick rather than a method, but it was taken up much later by others and shown to be very effective. Again, we shall not give references. The relevant papers are cited in a survey by MAGNUS [1981].

C. A Problem from the Foundations of Geometry

HILBERT [1901 and 1930] had shown that there exists a planar geometry in which all axioms of incidence and order are satisfied and in which the theorem of DESARGUES but not that of PAPPUS is universally valid. For his proof, he introduced what he called an ordered noncommutative number system (*Zahlsystem*), something which now would be called an ordered skew field or an ordered (associative but not commutative) division ring. Such a division ring must contain the field Q of rational numbers (which is generated by the unit element 1 of the ring), and the elements of Q must belong to the center of the ring, i.e., they must commute with all other elements.

MOUFANG [1937] notes that HILBERT's construction can be described as the embedding of the group ring $Q(G)$ of a particular group (with coefficients in Q) in a division ring. Indeed, if we take the group G generated by two elements s, t, and the defining relations

$$ts = \gamma st, \qquad \gamma s = s\gamma, \qquad \gamma t = t\gamma,$$

the element γ generates the center of G. It can be identified with an element $\neq 1$ of Q, and HILBERT chooses $\gamma = 2$. The group G is then ordered by the rules

$$\gamma^c s^n > s^m, \gamma^c t^n > t^m, \qquad \text{if } n < m, \text{ and for all } c \in \mathbb{Z};$$

$$\gamma^c s^k t^l > s^n t^m \qquad \text{if } k < n \text{ or if } k = n, l < m.$$

The group ring $Q(G)$ of G over the rational numbers Q can then be embedded in a division ring R with the general element

$$\rho = \sum_{k,l=-N}^{\infty} r_{kl} s^k t^l,$$

where N is finite but arbitrarily large and $r_{k,l}$ are rational numbers. We can order R by defining ρ to be positive if and only if the largest of the monomials $s^k t^l$ has a positive coefficient, and we also postulate that $\rho = 0$ implies that $r_{kl} = 0$ for all k, l. It should be noted that the ordering of the elements of the group G is such that for elements belonging to different cosets of the center, the larger one is always "infinitely larger" than the smaller one, which means that no positive multiple of the smaller one can exceed the larger one. This allows one to define an inverse for every ρ. We first choose the largest one of the terms

$$|r_{kl}|s^k t^l$$

which we denote by μ. Then $\rho\mu^{-1}$ has an expansion of the form $1 + \rho^*$, where 1 is infinitely larger than ρ^*, and we can define ρ^{-1} by

$$\rho^{-1} = \mu^{-1}\left(1 - \rho^* + \rho^{*2} \mp \cdots\right).$$

MOUFANG [1937] considers the case of a free metabelian group M with two generators a, b, i.e., the quotient group F/F'' of a free group with two free generators and shows that its group ring $Q(M)$ over the rationals Q can be embedded in an ordered division ring R. In her case, R contains an abelian subring R_0, consisting of the rational functions with rational coefficients of (countably) infinitely many indeterminates. The elements a, b then act, by conjugation, as automorphisms of R_0. It should be noted that the indeterminates generating R_0 cannot all be replaced by algebraically independent real numbers because the real numbers are subject to an archimedian ordering (every positive number will be exceeded in size by an integral multiple of 1). However, she shows that it is possible to reduce the number of these indeterminates to 1 by enlarging the center Q of R, using a sufficiently large subfield of the real numbers instead of Q. (The possibility that even in this case Q may suffice as the center of R has been left open by MOUFANG.)

The details of the construction of the ring R are too complex to be reproduced here. We just mention a byproduct of this construction which is easy to prove. The free semigroup generated by two elements a, b (but without a^{-1}, b^{-1}) can be imbedded faithfully in the group $M = F/F''$. This is equivalent to saying that in F/F'' any two distinct words in a, b with exclusively positive exponents for a and b define different elements of F/F''.

Apart from HILBERT's original construction (which, however, is not formulated as a problem in group theory), the paper by MOUFANG [1937] seems to be the first systematic investigation of this problem for nonabelian groups. It was followed, among others, by a paper by F. W. LEVI [1942a], which generalizes HILBERT's result by showing that every torsion-free group whose commutator subgroup is contained in the center can be ordered, by a very general investigation of *Lattice ordered groups* by G. BIRKHOFF [1942], and by two papers by B. H. NEUMANN [1949a and 1949b]. The second of these papers is particularly important because it contains a proof of the theorem that every ordered group can be embedded in an ordered division ring. It also contains an explicit proof of the fact that free groups can be ordered—something indicated by BIRKHOFF [1946]. Finally, it contains a bibliography which lists references to several papers on ordered algebraic structures. These provide good background information on the broader context within which the theory of ordered groups belongs. Both of NEUMANN's papers mention HILBERT's work as a motivation for his investigations. For the later literature, we refer again to BAUMSLAG [1974, Chapter 14].

D. Notes on Later Developments and Generalizations

So far in this chapter we have emphasized the influence of problems from other fields on the investigation of metabelian groups. If we look upon them from a purely group-theoretical point of view, they form one of many subclasses of the class of solvable groups, that is, of the class of groups G for which the sequence of derived (or higher commutator) groups $G, G', G'', \ldots$ terminates after finitely many steps with the unit element. The number of terms distinct from the unit element in this sequence is sometimes called the *step* (*Stufe*, in German) or else the *solvability length* of the solvable group, and the metabelian (but nonabelian) groups are then solvable of step 2. They are the subject of a large number of papers published (for infinite but finitely generated groups) mainly after 1950, and their residual properties, presentations, automorphisms, and other aspects have been investigated. Although MAL'CEV [1951] had shown that even the finitely generated groups may be infinitely related, the finitely generated groups can still be described in terms of the finitely generated group ring of the abelianized group as a ring R of operators acting on an R-module which, as an R-module, is finitely generated. We shall mention below another more sharply defined *finiteness condition* which is satisfied by finitely generated metabelian groups but, in general, not by finitely generated solvable groups of step 3.

A historical account of the theory of infinite solvable groups might be expected to start with the theory of infinite abelian groups. These we have excluded from the topics to be discussed in our book, and we mention only in passing that a general theory of abelian groups began to emerge with the papers by ULM [1933] on the classification of countable abelian groups and by PONTRIJAGIN [1934] on the theory of topological commutative groups. Here, however, difficulties arise only if we begin to consider infinitely generated groups, since the theory of finitely generated abelian groups had essentially been settled in 1879 by FROBENIUS and STICKELBERGER. Therefore, bypassing the theory of infinite abelian groups becomes perfectly legitimate if we confine ourselves to the theory of finitely generated solvable groups, although we may have to consider their infinitely generated and, obviously, solvable subgroups. This is what happened historically, but it turned out that additional finiteness conditions were needed in order to obtain nontrivial results.

EMMY NOETHER (1892 − 1935) had, in several papers, e.g., in NOETHER [1921], established the importance of so-called *chain conditions* for algebraic structures, in particular, for rings. In the case of groups, her *ascending chain condition* assumes the following form: A group G satisfies the maximum condition for an ascending chain of subgroups if every properly increasing sequence of subgroups

$$1 \subset G_1 \subset G_2 \subset G_3 \subset \cdots$$

of G must terminate after a finite number of steps with the group G. This condition is now usually abbreviated as *Max*. If we restrict the subgroups $G_1, G_2, \ldots$ to the class of normal subgroups, we shall call the same condition *Max-n*. (These notations are fairly standard now.)

In a sequence of five papers, K. A. HIRSCH [1938–1954] investigated finitely generated solvable groups which satisfy the condition Max. One of his results states that such a group G contains an ascending sequence of subgroups

$$H_0 = 1 \subset H_1 \subset H_2 \subset \cdots \subset H_n = G$$

such that each of these subgroups is normal in the following one and that the quotient groups H_{i+1}/H_i of these groups are cyclic for $i = 0, \ldots, n - 1$. P. HALL [1954a] coined the term *polycyclic* for these groups. The papers by MAL'CEV [1951], L. AUSLANDER [1967], and SWAN [1967] eventually established the remarkable fact that the class of polycyclic groups coincides with the class of all solvable groups of $n \times n$ matrices with entries from the ring of integers.

For the extensive literature on polycyclic groups (up to 1970), we refer to D. J. S. ROBINSON [1972]. There the reader will also find an account of the theory of solvable groups which satisfy the *Min condition*. By that, we mean, of course, that every properly descending sequence of subgroups of G, starting with G itself, terminates after finitely many terms with the unit element. The Min condition was introduced by ČERNIKOV [1940] in a broader context in a paper on infinite locally solvable groups. (Following KUROSH, a group is said to have a property *locally* if it is shared by all of its finitely generated subgroups.)

Of the many classes of solvable groups which are characterized by some type of finiteness condition (in a very broad sense) and whose theory is presented in the book by D. J. S. ROBINSON [1972], the so-called *nilpotent* groups will be discussed at some length in the next chapter. We conclude this chapter with a note on the role of the Max-n condition which shows its great importance for the theory of solvable groups and which also explains why, for instance, no systematic theory of groups of solvability length 3 exists. The following results are due to P. HALL [1954a]. In formulating them, we shall use some notations and concepts explained in the next chapter and also the concept of a variety of groups which will be explained in Chapter II.8.

Let G denote a finitely generated group. Define $\gamma_1 G = G$ and $\gamma_n G = (\gamma_{n-1}G, G)$ for $n > 1$ and $\delta_n G = (\gamma_n G, \gamma_n G)$. (The groups $\gamma_n G$ are simply the groups of the lower central series.) We call the γ_n, δ_n "subgroup functions". Using the commutator symbol $(\cdot, \cdot)$ we can define a large class of such subgroup functions, e.g., $(\delta_2 G, \gamma_1 G)$. We introduce a partial ordering for these functions by saying that $\phi \leqslant \psi$ if, for the subgroup functions ϕ, ψ, it is true for all groups G that ϕG is contained in ψG. We subdivide subgroup functions into two disjoint subclasses. The first subclass will contain all subgroup functions ϕ such that, for a suitably large n, $\delta_n \leqslant \phi$. The second subclass will contain all subgroup functions ψ such that $\psi \leqslant (\delta_2, \gamma_1)$. Now HALL proves: The finitely generated groups G belonging to the variety defined by $\phi G = 1$ (ϕ fixed), satisfy the condition Max-n, and there exist only countably many of them. But not all finitely generated groups G belonging to the variety $\psi G = 1$ (ψ fixed) satisfy Max-n, and there exist uncountably many such groups.

The metabelian groups are defined by $\delta_2 G = 1$, and, accordingly, they satisfy Max-n and there are countably many nonisomorphic finitely generated metabelian groups. On the other hand, the groups G for which $(\delta_2 G, \gamma_1 G) = 1$ do not always satisfy the condition Max-n. They are the groups for which the second commutator subgroup G'' belongs to the center of G. Here HALL proves a theorem which is even sharper than his general result. There exist uncountably many nonisomorphic such groups with two generators for which the center is any arbitrarily prescribed fixed nontrivial countable abelian group.

Our summary of the results of the paper by HALL [1954a] illustrates the increasing complexity of the terminology needed for this purpose, which is rather typical for many important papers published after 1945. A nearly historical account of the developments in the theory of infinite solvable groups up to 1970 may be found in Sections 65–69 of BAUMSLAG [1974]. Clearly, it would be futile to try to accommodate the contents of these 21 large-format pages covered with small print in our text.

Chapter II.7

Commutator Calculus
and the Lower Central Series

In 1933, there appeared a 66-page paper by P. HALL with the title *A contribution to the theory of groups of prime power order*. Its introduction describes it as "the first stages of an attempt to construct a systematic general theory of groups of prime power order." These groups are also called *p-groups*, where, if p is not specified as in 3-groups, etc., it is tacitly assumed that p denotes a prime number and that the order of the p-group is a power of p. Later, even infinite p-groups have been introduced as groups in which every element has an order which is a power of p (see, e.g., SANOV [1949]). HALL then briefly mentions earlier results on special p-groups, including the investigations of the American school and continues by describing his own approach saying:

> We have been guided by the idea that the structure of a group should be exhibited as far as possible by the inter-relations of its *characteristic subgroups*. A subgroup is characteristic, according to Frobenius, if it is transformed into itself by every automorphism of the group; thus characteristic subgroups represent what one would actually call *invariant* features of the group structure.

A footnote attached to the word "invariant" adds: "Were it not that the word invariant is used by some writers for the weaker notion of self-conjugate." (We note here that both the terms "invariant" and "self-conjugate" have since disappeared from the literature. They have been replaced by the rather nondescriptive word "normal".)

After mentioning some elementary methods of constructing new characteristic subgroups from given ones, in particular, the *product* and the *intersection* (*join* and *meet*, in HALL's vocabulary) of two given ones, HALL continues:

> Of great importance in the sequel is another mode of generating characteristic subgroups:— *commutation*.

Actually, the subgroups constructed by HALL using commutation are not only characteristic but even *fully invariant*. This concept was introduced in the same year by F. LEVI [1933] under the German name *vollinvariant*. It means that a fully invariant subgroup H is not only mapped onto itself by all automorphisms of the whole group G, but even mapped homomorphically into itself by every *endomorphism* of G, i.e., by every homomorphic mapping of G *into* itself (or *onto* a subgroup K of G). Nearly one-half of HALL's paper (Sections 2,3, pp. 42–73) deals largely with the construction of subgroups based on the use of commutation and on the relationship of commutation as a binary composition to the ordinary binary composition of group elements. The results obtained here are valid for all groups and, in particular, are completely independent of the theory of p-groups. The term *commutator calculus* for this type of result was introduced only much later by MAGNUS, KARRASS, and SOLITAR [1966] but is now (1980) standard (see, e.g., BAUMSLAG [1974, Chapter 8]).

We shall not be able to give a full account of the many results contained in the paper by HALL [1933]. A somewhat sketchy description of it, stressing the underlying ideas rather than the specific theorem, will be given below. The paper is the first of a long sequence of important contributions by P. HALL to group theory and to related topics. One of them has already been mentioned in Section II.6.D. Many of them deal with the theory of finite groups only and, for this reason, will not be mentioned in the present book. Apart from his own work, P. HALL has played a large role in the development of pure mathematics in England through the work of his numerous Ph.D. students. He himself became interested in group theory by reading the work of WILLIAM BURNSIDE (1852–1927) or, more precisely, his group-theoretical work, in particular, the *Theory of groups of finite order* (BURNSIDE [1911]) and the two papers on p-groups, BURNSIDE [1912 and 1913]. (BURNSIDE also worked in several other fields including hydrodynamics.) In the introduction to his 1933 paper, HALL quotes several results from these publications and mentions, in particular, that Theorem VII, p. 166 in BURNSIDE [1911] introduces, at least implicitly, the lower central series and proved that in a group of finite order it will terminate with the unit element if and only if the group is the direct product of groups of prime power order. Although one may call HALL (b. 1904) a student of BURNSIDE, this would not be correct in the sense in which this term is now commonly understood, where it presupposes direct communication through contemporary presence at the same university. At least since the end of the Nineteenth Century, the latter type of teacher-student relation has prevailed in the great majority of cases.

Now we have to introduce some notations. We shall try to keep technicalities to a minimum, and we shall use several notations which are more or less standard now but do not appear in HALL's 1933 paper, although some of them were introduced later by him. Also, we shall use the letters of the alphabet consistently as follows.

G will always be a group, and L, M, N will be normal subgroups of G. u, v, w, will denote elements of G. The letters a, b, c are reserved for generators of G.

For any two elements u, v, we define

$$(u, v) = u^{-1}v^{-1}uv, \qquad u^v = v^{-1}uv, \qquad u^{-v} = v^{-1}u^{-1}v.$$

Here (u, v) is the *commutator* of u and v. If L, M are any two normal (characteristic, fully invariant) subgroups of G, so are the products LM, their intersection $L \cap M$, and their *commutator* (L, M) which is defined as the group generated by all commutators (u, v) with $u \in L, v \in M$. Clearly, $(L, M) \subset L \cap M$.

If we consider commutation (i.e., the forming of the commutator) as a binary composition, the first difficulty which arises is one of notation, since commutation is a nonassociative operation. This means that there are $2(2n-3)!/[n!(n-2)!]$ possibilities of forming a composite (or, as we shall say, product) of n components (which we shall call factors or elements) which are given in a fixed order. (See, e.g., ERDÉLYI [1955]). They are determined by different *bracket arrangements*. For $n = 4$, these are

$$(((*,*),*),*); \; ((*,(*,*)),*); \; (*,(*,(*,*))); \; (*,((*,*),*)); \; ((*,*),(*,*)),$$

where the asterisks represent elements.

Independent of the various bracket arrangements, the number of constituents entering into the formation of a commutator is called its *weight W*. The number of brackets opening to the right in a commutator of weight W always equals $W-1$. If all of these brackets are on the left-hand side of the commutator, it is called a *simple commutator*. HALL writes

$$(u_1, u_2, \ldots, u_n) = ((u_1, u_2, \ldots, u_{n-1}), u_n)$$

for the simple commutator of weight n with *components* $u_1, \ldots, u_n$ (in this order). As mentioned above, we can substitute normal subgroups of G for the components of a commutator, the result again being a normal subgroup of G. If, in particular, we substitute G itself for the components of the simple commutators, we obtain the *groups of the lower central series* of G which were first introduced by FITE [1906]. They are denoted by G_n or by $\gamma_n G$ and are defined recursively by

$$G_1 = G, \; G_n = (G_{n-1}, G) \equiv \gamma_n G \qquad \text{for } n > 1.$$

[The notation $\gamma_n G$ is now (1980) the most widely accepted one. It was also introduced by P. HALL [1954]. HALL [1933] and later authors used G_n.] The term "lower central series" implies two properties of the sequence of the G_n: First, G_n/G_{n+1} is the center of G/G_{n+1}, and second, of all such descending sequences of normal subgroups of G, the lower central series is the one which is descending fastest. If Γ_n is the nth term in any other descending central series, Γ_n contains G_n. Of course, G_n is not only normal but even fully invariant in G. It is not at all necessary for G_{n+1} to be a proper subgroup of G_n; if not, then all later groups G_m with $m > n$ coincide with G_n, and this can happen even if $n = 1$. A group G for which $G_{n+1} = 1$ for a finite value of n, but $G_n \neq 1$ is called *nilpotent of class n*. This terminology has been taken over from the theory of Lie groups where it had been first used for the Lie algebra associated with the group. It does not yet appear in HALL [1933]. However, the term "class of a group" was coined by him.

The other sequence of commutator subgroups which plays an important role in HALL's paper is the *derived series*. Its terms $G^{(n)}$ (or $\delta_n G$) are defined recursively by

$$G^1 = (G, G), \qquad G^{(n)} = (G^{(n-1)}, G^{n-1}).$$

This sequence had, of course, been considered for a long time in the theory of finite groups. If $G^{(n)} = 1$ for a finite value of n, G is called *solvable*. Also, BURNSIDE's characterization of finite nilpotent groups (mentioned above) shows that solvable groups need not be nilpotent although nilpotent groups are solvable. A sharpened version of this statement is HALL's result which says that

$$G^{(n)} \subset \gamma_m G \text{ for } m = 2^n.$$

On the other hand, no $\gamma_n G$ is contained completely in G'' if G is a free nonabelian group. This is due to the fact that G'/G'' is in this and in many other cases an infinitely generated abelian group even if G is finitely generated. But, in this case, $\gamma_n G/\gamma_{n+1} G$ is a finitely generated group. More precisely, HALL showed that the simple commutators

$$(u_1, u_2, \ldots, u_n),$$

where $u_1, \ldots, u_n$ run independently through a full set of generators of G, generate a complete system of cosets of $\gamma_{n+1} G$ in $\gamma_n G$. (This result is not optimal; the exact upper bound for this number of generators is given by a formula due to WITT [1937] to be described later.) The abelian groups $\gamma_n G/\gamma_{n+1} G$ are invariants of G in the sense that they must be isomorphic for isomorphic groups. Since finitely generated abelian groups have been classified completely by FROBENIUS and STICKELBERGER [1879], this provides tests for the isomorphism of finitely generated groups. For $n = 2, 3$, REIDEMEISTER [1927] and also his Ph.D. student HERTHA ADELSBERGER [1930] had used this technique before. In her paper, G/G_4 appears in a slightly disguised form based on the Baker–Hausdorff formula which will be discussed briefly at the end of this chapter. HALL uses the technique largely in connection with the theory of p-groups which does not concern us here. Of his results relevant for groups in general, we mention the following:

Let L, M, N be any three normal subgroups of G. Define A, B, C as the simple commutators

$$A = (L, M, N), \qquad B = (M, N, L), \qquad C = (N, L, M).$$

Then each one of these groups is contained in the product (or the union, which is the same in this case) of the two other groups:

$$A \subset BC, \qquad B \subset CA, \qquad C \subset AB.$$

Note that $(L, M, N) = (M, L, N)$ and $BC = CB$. We cannot go into the applications of these formulas. Their proof is based on the following identities (i.e., relations valid for elements of any group):

$$(u, v)(v, u) = 1,$$

$$(v, vw) = (u, w)(u, v)(u, v, w), \qquad (*)$$

$$(uv, w) = (u, w)(u, w, v)(v, w).$$

These relations were supplemented later by HALL (unpublished) with the identity

$$((u, v), w^u) \cdot ((w, u), v^w) \cdot ((v, w), u^v) = 1 \qquad (**)$$

which is actually a symmetric form of an identity discovered earlier by WITT [1937] and used by him for the proof of a theorem to be formulated later in this chapter. The identities $(*)$, $(**)$ exhibit the very complicated nature of the relationship between commutation and the ordinary composition of group elements. The most sophisticated and also the most important of the relations of this type found by HALL can be described (almost) in HALL's words (although not with his notations) as follows.

For any two elements u and v of any group G, let the formally distinct complex (i.e., not necessarily simple) commutators of u and v be arranged in order of increasing weights, the order being otherwise arbitrary; thus

$$z_1, z_2, \ldots, z_i, \ldots .$$

Then there exists a series of integer-valued polynomials

$$f_1(n), f_2(n), \ldots, f_i(n), \ldots$$

(and $f_1(n) = f_2(n) = n$) all vanishing for $n = 0$ and such that the degree of $f_i(n)$ does not exceed the weight W_i of z_i in u and v, and with the property that

for all u and v and all positive integers n,

$$(uv)^n = z_1^{f_1(n)} z_2^{f_2(n)} \ldots z_i^{f_i(n)} \ldots ,$$

this equation being interpreted to mean that, if z_λ, $\lambda = \lambda(e)$, is the last term of the sequence $z_1, z_2, \ldots, z_i, \ldots$ which is of weight less than e in u and v, then

$$(uv)^n = z_1^{f_1(n)} z_2^{f_2(n)} \ldots z_\lambda^{f_\lambda(n)} \cdot \zeta_e(u, v), \qquad (***)$$

where ζ_e is an element (depending on u and v) of the group $\gamma_e(K)$, where K is the subgroup of G generated by u and v. Relation $(***)$ holds for all $e = 2, 3, \ldots$.

A particularly important special case of $(***)$ is obtained by taking $n = p$ (a prime number) and $e = p$. In this case, all of the values $f_1(p), \ldots, f_\lambda(p)$ are necessarily multiples of p. We shall return to this case at the end of this chapter in connection with BURNSIDE's problem.

The paper by P. HALL [1933] was reviewed by MAGNUS for the *Zentralblatt für Mathematik* (7, p. 291). Although the review emphasizes the theorems on p-groups, MAGNUS was particularly interested in formula $(***)$ (HALL's proof of which happens to be very difficult). It is a formula valid in a free group with free generators u, v and it actually has to be proved only in this case. Also, it has the character of an approximation formula with remainder term, and the first three terms can easily be computed explicitly, the result being, e.g.,

$$z_1 = u, \qquad z_2 = v, \qquad z_3 = (u, v),$$

$$f_1(n) = n, \qquad f_2(n) = n, \qquad f_3(n) = \frac{n(n-1)}{2}.$$

But the computation of higher terms becomes rapidly more difficult. Commutators become more and more unwieldy with increasing weight, and the long expression on the right-hand side of $(***)$ then reduces to the comparatively short word on the left-hand side because of cancellations of terms uu^{-1}, $u^{-1}u$, etc. Therefore, it would be desirable to get rid of the inverses. DYCK [1882] and others had tried to do that by introducing a third generator t and setting $uvt = 1$. But then one has to cancel the subwords uvt, tuv, vtu wherever they appear in a long word. However, for linear operators (matrices, integral operators, etc.) which frequently appear as elements of (multiplicative) groups, there exists a standard way of computing the inverse of an operator which in some sense is "close" to the identity operator. One simply uses the geometric series for $1/(1 + x)$. This trick presupposes that the operators appear as elements of a ring and that an infinite series makes sense in one way or another. Both conditions can easily be satisfied for free groups, using methods which had been established at the turn of the century by HILBERT in the construction of non-archimedian skew fields and by HENSEL in the construction of p-adic numbers. These somewhat vague considerations were made precise by MAGNUS [1935] as follows.

Let a_i, $i = 1, 2, \ldots$, be free generators of a free group F. We introduce a "free associative ring" R with generators s_i and with a unit element 1. Then R_0 consists of all finite linear combinations with integral coefficients of products

$$s_{i_1}^{e_1} s_{i_2}^{e_2} \ldots s_{i_n}^{e_n}, \qquad e_1, e_2, \ldots, e_n > 0, \qquad i_{k+1} \neq i_k. \tag{\#}$$

and these together with 1 form a basis of R_0 with coefficients in $\mathbb{Z}$. We extend R_0 to a ring R consisting of infinite sums of terms of this type. The elements s_i generate an ideal J in R. The quotient ring R/J is the ring $\mathbb{Z}$ of integers, and the quotient ring R/J^{d+1} has as basis elements 1 and those terms in (#) for which the sum

$$e_1 + e_2 + \cdots + e_n \leqslant d.$$

This sum will be called the *dimension* of the basis element, and also of any linear combination of basis elements of the same dimension. Any element of R which is of the form $1 + \omega$, where $\omega \in J$, has an inverse

$$(1 + \omega)^{-1} = 1 - \omega + \omega^2 - \omega^3 \pm \cdots .$$

The infinite series on the right-hand side is well defined since each power of ω contributes only finitely many terms of a fixed dimension and only finitely many powers contribute any terms at all of this type. Thus the mapping

$$a_i \to 1 + s_i$$

defines a mapping of the free group F into the multiplicative group of the ring R, and every word W in a_i has an expansion in R of the form

$$W \to 1 + S_1 + S_2 + \cdots + S_n + S_{n+1} + \cdots ,$$

where, on the right-hand side, the term S_n is (for $n = 1, 2, \ldots$) an element of dimension n. If all of the terms $S_1, S_2, \ldots$ are zero, we shall say that W has dimension ∞. Otherwise, we shall have a first term S_n which is not zero and then we shall call n the *dimension of W* and denote S_n by δW. Now MAGNUS [1935] proved:

(i)   The mapping $a_i \to 1 + s_i$ defines an isomorphic mapping of the free group F into the multiplicative group of R.

(ii) The elements of F with a dimension $\geqslant n$ form a fully invariant subgroup $D_n F$ of F. It is called the *nth dimension subgroup*.

(iii) The quotient groups $D_n F/D_{n+1} F$ are torsion-free abelian groups. They are finitely generated if F is finitely generated.

(iv)   The intersection of all of the $D_n F$ is the unit element.

(v) $D_n F$ contains the nth subgroup $\gamma_n F$ of the lower central series as a subgroup. If F is finitely generated, then $\gamma_n F$ is of finite index in $D_n F$.

The paper by MAGNUS [1935] also contains applications to other problems, in particular, to the computation of automorphism groups, to the isomorphism problem of one-relator groups, and to groups of prime power order. Now, there is a proof that finitely generated free groups are Hopfian, i.e., nonisomorphic with any one of their proper quotient groups. This result is implicit in the earlier paper by NIELSEN [1921] and in the paper by LEVI [1935] which is cited for this reason by MAGNUS [1935]. Finally, there is a finite-dimensional faithful matrix representation of $F/D_n F$.

Clearly, the ring R_0 could have been introduced as a subring of the group ring $\mathbb{Z}(F)$ of the free group F over the ring $\mathbb{Z}$ of integers as ring of coefficients. We then simply have

$$s_i = x_i - 1$$

and the s_i also generate an ideal J^* in $\mathbb{Z}(F)$. This ideal is now called (see Fox [1953]) the *fundamental ideal* or the *augmentation ideal* of F. It can also be defined as the ideal generated by all of the elements $g - 1$ of $\mathbb{Z}(F)$, where g is an arbitrary element of F, and this definition makes it possible to define the nth-dimension subgroup $D_n G$ of an arbitrary group G as the subgroup of elements h for which $h - 1$ belongs to J^{*n}, where J^* is generated by the elements $g - 1$, where $g \in G$. This construction which is due to Fox also makes it unnecessary to introduce infinite sums.

MAGNUS [1934] mentioned two unsolved problems. One of them is somewhat technical, arising from his application of the dimension subgroups to the theory of automorphisms of free and of one-relator groups. Without going into the details, we mention only that the automorphism group of a free group F on m free generators induces a group of automorphisms of the free abelian group $D_n F/D_{n+1} F$ which is a subgroup of the linear group $GL(d_n, \mathbb{Z})$ of $d_n \times d_n$ matrices with integral entries and determinant ± 1, where d_n denotes the rank of $D_n F/D_{n+1} F$. At the same time, this group is a matrix representation of the groups $GL(n, \mathbb{Z})$, and the question arises: What are its irreducible components? Partial answers have been given by WEVER [1949] and BURROW [1958].

The other problem actually was a conjecture which was proved shortly afterwards. It states that

$$D_n F = \gamma_n F$$

for all n and for all finitely generated free groups F.

The most noteworthy effect of this (rather plausible) conjecture was not its proof (which is difficult enough), but the introduction of Lie rings into combinatorial group theory as a useful tool. Before describing this development, we briefly mention the answer to the generalized form of this conjecture which was whether, perhaps, $D_n G = \gamma_n G$ for *all groups*. The answer is "no" according to RIPS [1972], although in many cases it is "yes". For partial answers to, and ramifications of, this question, see SANDLING [1972].

The first proof that $D_n F = \gamma_n F$ was given by GRÜN [1937]. It uses matrix representations for $F/\gamma_n F$ and is not easy to follow. Also, it seems to have left no trace in the literature. The next proof is based on an approach proposed by MAGNUS [1937a] who formulated the following theorem as a prerequisite for the proof of $D_n F = \gamma_n F$:

Define a *free Lie ring* Λ_0 on generators σ_i $(i = 1, 2, \ldots)$, as follows: In Λ_0, there exist two binary operations, called addition and multiplication. The elements of Λ_0 form an abelian group under addition. Multiplication and addition are connected by the distributive laws, but multiplication is neither commutative nor associative. Denoting the product of any two elements ϕ, ψ of Λ_0 by $[\phi\psi]$, we have, instead, for any two elements ϕ, ψ,

$$[\phi\psi] + [\psi\phi] = 0,$$

and for any three elements ϕ, ψ, χ, the relation

$$[[\phi\psi]\chi] + [[\chi\phi]\psi] + [[\psi\chi]\phi] = 0$$

which is called the *Jacobi identity*. We shall say that an element ϕ, which has been derived from the σ_i by multiplication only and contains d terms which are taken from the set of the σ_i (possibly with repetitions), has *degree d*, and we shall say the same about any linear combination with integral coefficients of such elements. Now we define a mapping μ of the elements of Λ_0 onto certain elements of R_0 which we shall call *Lie elements* of R_0. This mapping is defined as:

(i) $\sigma_i \to s_i$;
(ii) if ϕ, ψ are in Λ_0, and $\phi \to u$, $\psi \to v$, where u, v are in R_0, then

$$\phi \pm \psi \to u \pm v, \qquad [\phi\psi] \to uv - vu.$$

The expression $uv - vu$ is also denoted by $[u, v]$ and is called the *bracket product* or (in the older literature) the *alternant* of u and v.

Now the theorem in question states: *The mapping μ is a one-to-one mapping of Λ_0 into R_0*. It was proved by MAGNUS [1937b], who also showed that the additive group of Lie elements of degree (or dimension) d is the direct sum of infinite cyclic groups and who gave a recursive construction of a set of linearly independent (over the integers) basis elements for the Lie elements of degree d in the case of a finitely generated free Lie ring Λ_0. Together with the theorem that $D_n F = \gamma_n F$, this provides the construction of a basis for the free abelian group $\gamma_n F / \gamma_{n+1} F$ in the case where F is finitely generated. M. HALL [1950b] derived an equivalent construction which is based on the method used by P. HALL [1933] to derive formula $(***)$ given earlier in this chapter.

To prove that $D_d F = \gamma_d F$ for all d, it is necessary to prove also that the elements $S_d = \delta W$ which appear in the representation of a word $W \in F$ of dimension d as an element of R are the Lie elements of dimension d in R_0. There is a gap in the proof of this fact as given by MAGNUS [1937b]. To fill this gap, one needs an identity of type $(**)$ mentioned earlier. Together with a complete proof of $D_d F = \gamma_d F$, it was supplied by WITT [1937], whose paper is important for several reasons.

We may look upon the construction described above which maps the free Lie ring Λ_0 into the associative ring R_0 as a faithful *representation of Λ_0 in R_0*. It is characterized by the fact that every element of Λ_0 is the image of exactly one element of R_0 under a mapping μ, and that the product $[\phi, \psi]$ of two elements in Λ_0 is the image of the element $uv - vu$ in R_0, where $\phi = \mu u$ and $\psi = \mu v$. In the case considered so far, the rings Λ_0 and R_0 have a basis and every ring element is a unique linear combination of basis elements with integral coefficients. It is easy to define rings with the same property and with elements from an arbitrary fixed field K acting as coefficients, and we may then also consider general Lie rings Λ with coefficients in K as quotient rings of a free Lie ring. Now WITT [1937] proved the theorem:

Let Λ be an arbitrary Lie ring with coefficients in a fixed field K. Then there exists exactly one associative ring A, also with coefficients from K, such that A contains a faithful representation of Λ and is generated by those of its elements which represent elements of Λ. In the case where Λ is free and freely generated by certain elements σ_i, the ring A is the free associative ring R_0 with coefficients in K which is generated by those of its elements s_i which represent σ_i of Λ.

The algebra A is now called the *universal enveloping algebra* of the Lie algebra Λ, and WITT's theorem stated above is called the *Poincaré–Birkhoff–Witt theorem* by JACOBSON [1961, p. 156]. POINCARÉ [1899] and BIRKHOFF [1937] both constructed associative algebras whose Lie elements form a given Lie algebra. (For nilpotent Lie algebras, this is a difficult problem even in the case of a finite basis.)

Another important result obtained by WITT [1937] is an explicit formula for the number of linearly independent Lie elements of dimension d in a finitely generated free associative algebra. This formula also expresses the rank of the free abelian group $\gamma_d F / \gamma_{d+1} F$ of a finitely generated free group F.

The next application of the theory of the lower central series to groups given in terms of a presentation is contained in MAGNUS [1939c]. It starts with the observation that the lower central series is not needed to prove that finitely generated free groups are Hopfian since this could also be derived from the fact that, given an element $g \neq 1$ of a free group F, there exists a subgroup of F whose index is a power of 2 and which does not contain g. A systematic investigation of this and of related more general questions is due to M. HALL [1949a and 1949b]. The main topic of the paper by MAGNUS [1939c] is the proof of the following theorem which uses only the fact that the groups $\gamma_n F$ intersect in the unit element and which states:

If a group G is presented in terms of $m + r$ generators $a_1, \ldots, a_m, b_1, \ldots, b_r$, and of r defining relations, and if $b_1 = \cdots = b_r = 1$ in G, then G is a free group freely generated by $a_1, \ldots, a_m$.

This result is not as obvious as it appears to be. A brief proof using entirely different powerful tools which were developed much later is due to STAMMBACH [1967]. The many uses of the lower central series in combinatorial group theory which were developed later are summarized in Chapters VII and VIII of BAUMSLAG [1974]. We cannot report here on these developments, but we shall briefly describe the beginnings of a theory connecting Lie rings and residually nilpotent groups and some of the results derived from P. HALL's formula $(\ast\ast\ast)$ which culminate in the papers by KOSTRIKIN [1957 and 1958].

MAGNUS [1940] showed that the lower central series of a group G defines a Lie algebra associated with G. He also introduced a formula found by BAKER [1904] and HAUSDORFF [1906] as a computational aid in the commutator calculus. This formula expresses the following fact: Let s_1, s_2 be generators of a free associative ring R with coefficients in the ring Q of rational numbers. Define the exponential function $\exp s$ by

$$\exp s = \sum_{n=0}^{\infty} \frac{s^n}{n!} \, .$$

Then, if we set

$$\exp s_1 \, \exp s_2 = \exp s_3,$$

s_3 is a Lie element of R. MAGNUS became acquainted with this formula through a conversation with REIDEMEISTER whose paper of 1927 uses an approach suggested by HAUSDORFF's paper but does not cite it. (We shall have to say more about the Baker–Hausdorff formula at the end of this chapter.)

Simultaneously with MAGNUS, ZASSENHAUS [1940] constructed for any given p-group an associated Lie ring with coefficients in a ring of characteristic p. The method is not derivable from the one used by MAGNUS. It is based on his extensive investigation of Lie rings with prime number characteristic (ZASSENHAUS [1939]). A systematic theory of the relation between the lower central series of a group and the associated Lie ring was developed by LAZARD [1954]. His results include all previous ones as special cases. Also, his paper contains numerous applications (e.g., a new proof of one of the results in MAL'CEV [1949]) and examples.

We return now to P. HALL's formula ($*\,*\,*$) and its applications. HALL himself had used it mainly in the case where the exponent n is a power q of a prime number p in order to characterize p-groups in which the q-th powers themselves form a group. This means that for every fixed power of p, the product of the q-th powers of two group elements is again the q-th power of an element of the group. His formula shows that every p-group of class less than p has this property. MAGNUS [1937b] had shown that the coincidence of the groups of the lower central series with the dimension subgroups of a free group leads to an easy proof of HALL's formula. We shall now sketch its bearing on *Burnside's Problem* which is stated as follows: Let $B(n, e)$ be the group with n generators and with the (possibly countably infinite) set of defining relators which state that for any element x in B, $x^e = 1$. For which values of $n \geqslant 2$ and e will B be a finite group? We mentioned the early history of this problem in Chapter I.6. Now we shall deal only with the case where $e = p$ is a prime number $\geqslant 5$. In order to explain the role of HALL's formula for this problem, we shall rewrite it in the following form for $n = p$:

$$(a_1 a_2)^p = a_1^p a_2^p c_2^p \ldots c_{p-1}^p \zeta_p,$$

where a_1, a_2 are generators of a free group F, c_ν, $\nu = 2,\ldots,p-1$ is an element of $\gamma_\nu F$, and ζ_p is an element of $\gamma_p F$. Using the method of the dimension subgroups of F, we know that $\delta\zeta_p$ is a homogeneous Lie element of exact weight p in s_1, s_2. It is not determined uniquely as such, but it is determined $\bmod\, p$, i.e., up to p times a linear combination of Lie elements of weight p. It is easily seen that in our notation used for the theory of dimension subgroups,

$$\delta\zeta_p \equiv (s_1 + s_2)^p - s_1^p - s_2^p \bmod p.$$

That the right-hand side of this congruence is indeed ($\bmod\, p$) a homogeneous Lie element of degree p in R_0 was first proved by JACOBSON [1937] in a completely different context. It was rediscovered by ZASSENHAUS [1939], who gives credit to ARTIN for his explicit formula for $\delta\zeta_p$. The relevance of this relation for Burnside's problem has been noted by ZASSENHAUS [1940] and MAGNUS [1940], and its systematic application to that problem starts with a paper by GRÜN [1940], which contains numerous generalizations and extensions of the group-theoretical equivalent of the formula for $\delta\zeta_p$.

This "equivalent" is derived as follows: There exists a well-defined word ω_p in a_1, a_2 which is a product of commutators (simple commutators, if we wish) of exact weight p in a_1, a_2 and which represents an element belonging to $\gamma_p F$ but not to $\gamma_{p+1} F$ such that

$$\delta\omega_p = \delta\zeta_p.$$

In addition, ω_p is not a pth power of any element of F. However, if we now consider a_1, a_2 as generators of a group $B(2, p)$, then, in this group, ω_p defines an element not only of $\gamma_p B$ but even of $\gamma_{p+1} B$. Also, any commutator of weight $m+1$ with ω_p as a component will define an element of $\gamma_{m+p+1} B$. Since, in B, any relation which holds for the generators holds for any pair of elements, we can obtain many more relations stating that in B a commutator belongs to a higher group of the lower central series than is indicated by its weight. Suppose now that we are able to derive sufficiently many relations of this type so that we can prove that for a sufficiently large m and a given p,

$$\gamma_m B(2, p) \in \gamma_{m+1} B(2, p).$$

We then have not solved the Burnside problem for two-generator groups with exponent p, but we have (for these groups) solved what is called the *restricted Burnside problem* (a term introduced by MAGNUS [1950]). This is the following question: Does there exist a maximal finite p-group $B_0(n, p)$ such that all finite p-groups with at most n generators and with elements of order p only are quotient groups of $B_0(n, p)$?

All of these deliberations are stated in precise technical language in the above-mentioned paper by GRÜN [1940]. Skipping nearly a dozen papers on the restricted Burnside problem (reviews of which are printed in BAUMSLAG [1974, Section 7B]), we now turn to the paper by KOSTRIKIN [1958], who proved that the maximal finite group $B_0(n, p)$ exists for all finite n and all prime numbers p and who derived this theorem from the following theorem (also proved by him) on Lie rings.

Let $\Lambda(n, p)$ be the Lie ring with n generators and with coefficients from the Galois field of p elements (i.e., with integer coefficients $\bmod p$). Assume that in $\Lambda(n, p)$ the $(p-1)$-st Engel condition is satisfied which states that for any two elements ϕ, ψ of $\Lambda(n, p)$, the product with $p-1$ factors ψ:

$$[[\cdots[[\phi, \psi]\psi]\cdots]\psi] = 0.$$

Then $\Lambda(n, p)$ is nilpotent, i.e., every product of m factors in Λ is zero for a fixed sufficiently large m.

To understand the appearance of the Engel condition, one has to go back to the explicit form of $\delta\zeta_p$. We shall not explain the details of this argument which are rather technical. A coherent systematic presentation and proof of the basic facts of the theory up to 1960 (excluding, however, the advanced results of LAZARD and KOSTRIKIN) may be found in Chapter 5 of MAGNUS, KARRASS and SOLITAR [1966].

Almost simultaneously with the proof of KOSTRIKIN that the restricted Burnside problem has an affirmative answer, there appeared an announcement by P. S. NOVIKOV that, for all odd exponents $e > 72$ and for at least two generators, the full Burnside problem has a negative solution. This announcement turned out to be somewhat premature though it has not been refuted. We still do not even know whether $B(2,5)$ is infinite or not, although we know (M. HALL [1958]) that $B(n,6)$ is finite. But the safely established result that $B(n,e)$ is infinite for all odd exponents $e \geqslant 4381$ (subsequently reduced to $e \geqslant 665$) was obtained only later with the very substantial collaboration of S. I. ADJAN (with NOVIKOV) and has been presented in detail by ADJAN [1975], together with solutions of the word-and-conjugacy problem for these groups.

A comprehensive bibliography listing papers connected with the Burnside problem has been compiled by M. F. NEWMAN and published in MENNICKE, [1980, pp. 255–271]. Very much like "Fermat's last theorem" in number theory, Burnside's problem has acted as a catalyst for research in group theory. The fascination exerted by a problem with an extremely simple formulation which then turns out to be extremely difficult has something irresistible about it to the mind of the mathematician.

We conclude our account of the Burnside problem with the remark that a combination of the theorems of KOSTRIKIN and of NOVIKOV and ADJAN immediately provides the following result:

For all prime numbers $p > 665$, there exists a finitely generated simple infinite group without subgroups of finite index in which every element $\neq 1$ has order p. This is only one of several theorems appearing in the work of NOVIKOV and ADJAN which illustrate the fact that even finitely generated infinite groups need not share any simple properties with finite groups.

The theory of Lie rings, whose entry into combinatorial group theory we have sketched above, is itself an offshoot of a very important part of group theory. It began under the name *theory of groups of transformations* or *theory of continuous groups* (where *continuous* is a translation of the German word *continuierlich* or, after 1900, *kontinuierlich* rather than the German word *stetig* which has the same translation). Its present universally accepted name is *theory of Lie groups* in honor of its founder MARIUS SOPHUS LIE (1842–1899), who is known as SOPHUS LIE. The ideas, problems, methods, and results of this theory have entered not only into many parts of mathematics (above all analysis, geometry, and, since L. E. DICKSON [1901 and 1905], even the theory of finite groups), but also into Twentieth Century theoretical physics. It is clearly impossible to give here anything even remotely resembling a sketch of the history of this subject, and we shall now contribute only a few bits of information relating to the history of the theory of Lie rings, starting with the remark that the term "Engel condition" refers to FRIEDRICH ENGEL (1860–1941), an early and important collaborator of LIE.

LIE had constructed a Lie algebra (i.e., a Lie ring with finite basis and coefficients from the field of real or of complex numbers) with every transformation group which satisfies certain finiteness (number of parameters) and regularity (differentiability) conditions. Immediately, the question arose whether it would be possible to find a transformation group belonging to any given Lie algebra and to determine, up to group-theoretical isomorphism and topological homeomorphism (as we might say today), all such groups. This question led to a thorough investigation of the classification of Lie algebras and of linear groups, i.e., of groups of finite-dimensional invertible matrices which are the most accessible transformation groups. Suppose that we have a Lie algebra Λ with basis elements $\beta_1,\ldots,\beta_n$ and that we can find matrices $B_1,\ldots,B_n$ such that, in the associative algebra A with basis elements $B_1,\ldots,B_n$, the mapping $B_\nu \to \beta_\nu$ ($\nu = 1,\ldots,n$), and the bracket multiplication in A leads to a faithful representation of Λ in A. Now let $t_1,\ldots,t_n$ be n (real or complex) parameters and form the matrices

$$\exp(t_1 B_1 + \cdots + t_n B_n) = M(t_1,\ldots,t_n).$$

Here the exponential function is defined as an infinite series which is easily seen to be convergent for all values of the parameters. Its role is based on the fact that the exponential function is an eigenfunction of the differential operator; differentiation of M with respect to t_ν means multiplication by B_ν, at least at $t_1 = \cdots = t_n = 0$. From this, it follows that the exponential matrix function M above generates a group depending on n parameters t_ν with the given Lie algebra as *its* Lie algebra.

At this point, there enters a discovery made by CAMPBELL [1898] which a little later led to the formula of BAKER and HAUSDORFF. CAMPBELL showed that, at least for sufficiently small values of the $|t_\nu|$ and the $|\tau_\nu|$,

$$M(t_1,\ldots,t_n)\, M(\tau_1,\ldots,\tau_n)$$

is again a matrix which can be written as the exponential function of a Lie element (depending on the t_ν, τ_ν) in A. The Baker–Hausdorff formula is an abstract construction which avoids the problems of convergence (expressed by the term "sufficiently small value") by using formal power series in noncommuting indeterminates instead of matrices. It establishes an important fact: Not only is it possible to embed a Lie ring in an associative ring by using bracket multiplication, but, conversely, it is also possible, at least in a free Lie ring with coefficients in a field of characteristic zero, to introduce an associative multiplication. The Poincaré–Birkhoff–Witt theorem formulated above achieves, of course, much more. It actually constructs the minimal associative ring from which the given Lie ring (now defined with coefficients from an arbitrary field and not necessarily free) can be derived by bracket multiplication. This theorem has now replaced the Campbell–Baker–Hausdorff formula in the theory of Lie groups (see, e.g., JACOBSON [1961] or HUMPHREYS [1972]), and it was this type of application which motivated the paper by BIRKHOFF [1937] which was submitted for publication only a few weeks before WITT's paper, on the other side of the Atlantic Ocean. The remarkable thing here is not so much the coincidence in time, but the fact that BIRKHOFF states as "Theorem 3" that:

> The free Lie algebra with n generators is isomorphic with the free algebra of alternants involving n symbols.

He continues: "In other words, the identities of Lie–Jacobi imply all other identities true for alternants." The author's emphasis on this result has nothing to do with the main purpose of the paper which, as its title states, is the representability of Lie algebras and Lie groups by matrices. Rather, it is due to the fact that BIRKHOFF was a prominent representative of a general trend towards abstraction and generalization which emerged after the World War I and was particularly strong in algebra where its most influential author was probably EMMY NOETHER. This remark about BIRKHOFF also applies to JACOBSON, whose 1937 paper has been mentioned earlier in this chapter. Apart from the formula quoted there, it introduces *derivation* as a general algebraic concept whose importance is not confined to analysis, and it is one of several papers by the author which helped to establish the theory of Lie algebras as an autonomous field of mathematical research.

We conclude this chapter by noting that the main topic of the paper by BIRKHOFF [1937] also appears in ADO [1936]. This paper is in Russian and mentioned briefly by BIRKHOFF. [A continuation (ADO [1947]) has been translated into English.] Nevertheless, the authors clearly worked independently, and the near simultaneous appearance of their papers documents the fact that, in spite of increasing specialization, there exists an international consensus about the importance of some problems even though they are of a rather technical nature and not famous in the sense of some conjectures or of HILBERT's problems.

Chapter II.8

Varieties of Groups

Roughly speaking, a variety of groups is a class of groups in which certain relations called *laws* or *rules* or *identities* are universally valid. The most widely investigated variety of groups is that of abelian groups in which the commutative law is universally valid or, in other words, in which for arbitrary elements x_1, x_2 of the group, the identity

$$(x_1, x_2) = 1$$

holds, where, as usual, (x_1, x_2) is defined as the commutator

$$x_1^{-1} x_2^{-1} x_1 x_2$$

of x_1 and x_2. Similarly, the Burnside variety of exponent e is defined as the class of groups in which, for all elements x, the relation

$$x^e = 1$$

holds. The metabelian groups of Chapter II.6 can be defined as the variety of groups defined by the identity

$$((x_1, x_2), (x_3, x_4)) = 1,$$

and the nilpotent groups of class c discussed in Chapter II.7 form the variety defined by the identity

$$(x_1, x_2, \ldots, x_c, x_{c+1}) = 1,$$

where the parentheses ( ) denote the simple $(c + 1)$-fold commutator defined in Chapter II.7.

All of these varieties are defined by a single law or identity, but of course there is no difficulty in defining varieties by using more than one law, e.g., the variety of metabelian groups which are nilpotent of class c. All of the particular varieties mentioned so far have been discussed in earlier chapters.

BIRKHOFF [1935] formulated in great generality the concept of varieties not only of groups but of other algebraic structures as well and proved some general theorems about them. The beginning of a systematic theory of varieties of groups is marked by a paper by B. H. NEUMANN [1937a] which appeared in a German journal but in English. In earlier chapters we have cited papers by B. H. NEUMANN with dates of publication both before and after 1937, but this particular paper is the first of his many major contributions to combinatorial group theory. As it happens, it is also part of the Ph.D. thesis with which he earned his second doctorate awarded to him by the University of Cambridge in England where he had gone as a refugee from Germany in 1933. (The thesis for his first doctorate, from the University of Berlin, had appeared in 1932.)

Although NEUMANN [1937a] did not use the term *variety of groups* (it was coined by P. HALL in 1949), his terminology closely follows that used then in algebraic geometry. He also uses the term *endomorphism* (meaning a homomorphic mapping of a group onto one of its subgroups) which had been introduced by P. HALL. He then begins with the following definition.

Let F_n be a free group with free generators $x_1, x_2, \ldots, x_n$. These will be called *variables* and any word $W(x_1, \ldots, x_n)$ in these generators will be called a *function*. Let G be an arbitrary group and let $(a_1, \ldots, a_n)$ be an arbitrary n-tuple of elements of G. This n-tuple will be called a *point*, and the element $W(a_1, \ldots, a_n)$ of G will be called the *value* of the function W at this point. NEUMANN then proves: The values of a function or, more generally, of a set of functions (not necessarily in the same number of variables) at all the points of G generate a fully invariant subgroup of G. NEUMANN calls such a subgroup a *word subgroup*. It is now called a *verbal subgroup* and we shall use this term henceforth. A group G may have fully invariant subgroups which are not verbal subgroups, e.g., the subgroup generated by all elements whose order is a fixed prime number p, but in a free group, all fully invariant subgroups are indeed verbal subgroups.

A verbal subgroup is determined uniquely by a set Σ of functions W_σ of variables $x_1, x_2, \ldots$, where, of course, only finitely many variables occur in each W_σ and where, necessarily, the number of the W_σ is countable. Let F be a free group and let $\Omega_\Sigma(F)$ be the verbal subgroup of F generated by the W_σ. Then

$$F/\Omega_\Sigma(F) = V(F)$$

is called a *reduced free* or *relatively free* group defined by the *laws* $W_\sigma = 1$. Such a group can also be defined by the fact that it possesses a set of generators such that every relator of these generators is a law for the group. The class of relatively free groups with the laws $W_\sigma = 1$, together with all of their quotient groups, forms the *variety* of groups defined by the laws $W_\sigma = 1$. If the free group F from which we started has n free generators (n

finite), we denote the resulting relatively free group of the variety by $V_n(F)$ and call n its rank. It can be shown that $V_n(F)$ can always be defined by functions W_σ in at most n variables.

There exist two obvious varieties. One of them consists only of the groups of order 1 and can be defined by the law $x = 1$. The other consists of the variety of all groups. A given group G can belong to different varieties. However, if it contains a free subgroup of rank 2, it can only belong to the variety of all groups. Probably the most important aspects of the paper by NEUMANN [1937a] can be summarized in the following statements:

(i)   Every group G uniquely defines a particular variety which we shall denote by $\phi(G)$. Its finitely generated relatively free groups will be denoted by $\phi_n(G)$.

(ii) If G is finite, then the groups $\phi_n(G)$ are also finite, and they are subgroups of the direct product of $N_n(G)$ groups isomorphic with G, where $N_n(G) \leqslant |G|^n$ with $|G| = $ order of G.

(iii) There exists an algorithm which allows one to compute $\phi_n(G)$ for a finite group G directly from G.

NEUMANN illustrates (iii) with explicit examples. He also cites a paper by P. HALL [1936] which starts with an entirely different problem, uses nothing from the theory of varieties, and yet contains an interesting result relevant to the computation of $\phi_n(G)$ in the case where G is a simple group of finite order.

In this case, HALL shows how to compute the maximal number $d_n(G)$ defined by the fact that a direct product of $d_n(G)$ groups isomorphic with G can be generated by n elements. For $n = 2$ and the simple groups G of order 60, 168, 360, 660, and 1092, he gives numerical values for $d_2(G)$. Clearly,

$$d_2(G) \leqslant N_2(G).$$

After the paper by B. H. NEUMANN [1937a], the theory of varieties of groups remained nearly dormant for a long time. The next two papers on this subject were those of LEVI [1942] and WEVER [1950]. Chapter 6 of BAUMSLAG [1974] lists 108 papers on varieties of groups which were reviewed in the years 1954–1970. They include publications by B. H. NEUMANN, HANNA NEUMANN (1914–1971), and their son P. M. NEUMANN. The monograph by HANNA NEUMANN [1967] is still the standard work on the subject. It has been particularly influential through the many problems formulated by the author which stimulated much of the later research. Up to 1973, the papers dealing with these problems have been summarized by KOVÁCS and NEUMANN [1973]. The bibliography of this survey alone has 67 entries.

Here we shall briefly describe only two topics from the theory of varieties. One of them is a theorem which appeared briefly in Chapter I.6

about the construction of group extensions. It was discovered by
KALOUJNINE and KRASNER in 1948 [1951] well before the proliferation of
papers on varieties, but providing an important tool for their investigation
and for the construction of groups with certain properties. KALOUJNINE and
KRASNER constructed what they called the *complete product* of a set
$\Gamma_1, \Gamma_2, \ldots, \Gamma_s$ of permutation groups and they showed that, in the case $s = 2$,
their product of two groups A and B produces a group which contains all
extensions of A by B (i.e., all groups G with a normal subgroup isomorphic
to A and $G/A \simeq B$) as subgroups. The term *wreath product* is derived from
the German word *Gruppenkranz* which was coined by POLYA [1937]. The
English term appears in print for the first time in B. H. NEUMANN [1956].
Following the notation and definition of BAUMSLAG [1959], we construct a
group $A \wr B$ (read: A wreath B) of two groups A and B as follows:

Let A_b be an isomorphic copy of A indexed by the elements b of B. Form
the direct product K of the groups A_b and let a_b be an arbitrary element of
A_b. Then, for any element c of B, the action of c on K is defined by

$$c^{-1}a_b c = a_{bc}.$$

This defines $A \wr B$ as an extension of the group K by a group isomorphic
with B.

It has been observed by several authors (see, e.g., H. NEUMANN [1967, p.
46]) and mentioned in Chapter I.6 that the construction of the wreath
product is nothing new. Indeed, it can be described in elementary terms
roughly as follows: Take the representation of B as a right regular group of
permutations and replace the entries $\neq 0$ in each of these permutations in
all possible ways by elements of A. The subgroup of diagonal matrices in
the resulting group is clearly isomorphic to K. The monomial group
representations are a first step in the direction of this construction. Nevertheless, we believe that the uses made of it by KALOUJNINE and KRASNER
justify our assigning priority to these authors. In the theory of varieties, the
wreath product serves several purposes, the most important of which is
probably the construction of the product UV of two varieties U and V. This
is the variety of groups G having a normal subgroup N belonging to U such
that G/N belongs to V. For all of these topics, including historical footnotes, definitions, and generalizations, see H. NEUMANN [1967, Chapter 2
(Product Varieties)].

The second topic we want to mention may be described as a fundamental
problem in the theory of varieties. It is the analog of a famous problem from
the classical theory of algebraic invariants which had been answered by
HILBERT [1880] and has an equally simple formulation: Can every variety be
defined by a finite number of laws? For the varieties defined by finite
groups, this question had already been raised by B. H. NEUMANN [1937a].
For these, it was answered in the affirmative by SHEILA OATES and M. B.

POWELL [1964]. (Of course, one has to consider here the varieties of infinite rank. Trivially, a finitely generated relatively free group defined by a finite group, being finite itself, can be defined by finitely many laws.) But for varieties containing infinite finitely generated groups, the problem is much more difficult. Partial results were listed on p. 22 in H. NEUMANN [1967]. The full problem was solved three years later. Within six months, in 1970, OL'ŠANSKIĬ, VAUGHAN-LEE, and ADIAN gave examples of varieties which can be defined only by infinitely many laws. The example given by ADIAN is particularly remarkable because it involves only laws in two variables. On the other hand, it is based on the very difficult results which he and P. S. NOVIKOV had obtained in their proof that there exist Burnside groups of infinite order.

As an immediate consequence of the existence of these examples, we have the result that there exist uncountably many different varieties of groups.

We cannot explain why, after the publication of the paper by B. H. NEUMANN [1937a], it took such a long time for a revival of interest in the theory of varieties of groups. We have a similar but not nearly so striking example from the theory of associative rings. In the same year, 1937, the doctoral thesis of W. WAGNER, DEHN's last Ph.D. student in Germany appeared. The topic of this important paper might very well be called an investigation of varieties of rings. It is now called *rings with polynomial identities* or simply *P.I.-rings* (an abbreviation which the uninitiated sometimes interpret to mean "principal ideal rings"). WAGNER's paper seems to have had no direct influence on the development of this field, although it was fully recognized by KAPLANSKY [1948] in his first survey of the results obtained at that time.

Chapter II.9

Topological Properties of Groups and Group Extensions

In Chapter I.5 we described the first productive uses made by DEHN and others of CAYLEY's discovery of the graph of a group. In (rather vague) general terms, this may be described as a method of associating a topological cell complex (in the case of the graph, a 1-complex) with the group and deriving results about the group from this association. The first progress beyond the methods used by DEHN is probably the connection between covering spaces and subgroups of the fundamental group of the base space which had been utilized by REIDEMEISTER [1932b]. An important and very productive generalization was introduced by VAN KAMPEN [1933a and 1933b], who uses 2-complexes. His ideas, which became fully effective only 30 years later, have been explained by LYNDON and SCHUPP [1977, pp. 151 and 236]. Chapters 3 and 5 of their work deal with these geometric methods, explaining all technicalities and providing some historical information. The title of Chapter 5 is *Small cancellation theory*. LYNDON and SCHUPP treat it by geometric methods. It started as a purely algebraic theory in a paper by TARTAKOVSKII [1949], who provided algorithms for the solution of word problems in a large class of groups which have presentations in which only sufficiently small parts of the relators can cancel each other. These algorithms are, in part, highly complex. Still using purely algebraic arguments, GREENDLINGER [1960a and 1960b] introduced classes of groups with similar assumptions about the relators in which the word and conjugacy problem can be solved by applying an algorithm of the very simple type which DEHN [1912] had used to solve these problems for the fundamental groups of closed two-dimensional manifolds (see Chapter I.5). DEHN, too, had used arguments of a geometric type, but the methods employed by LYNDON and SCHUPP [1977] (which, in large part, are due to the authors) are of a much more sophisticated and, unavoidably, more complex nature and cannot be presented here.

The possibility of associating a simplicial cell complex with a group also has other consequences of a purely group-theoretical nature. To every such complex there belong several groups. Of these, the role of the fundamental (or first homotopy) group has been mentioned many times in earlier chapters. But two infinite sequences of abelian groups also belong to such a complex, and are invariants of the space which is defined by it. They are the *homology groups* (formerly called *Betti groups*) and the *cohomology groups*. In papers by several authors which appeared in the 1940's, it was shown that the association of cell complexes and groups leads to a purely algebraic definition of homology and cohomology groups belonging to any given group G which are invariants of G (i.e., are isomorphic for groups isomorphic to G). Also, FREUDENTHAL [1931] had introduced the concept of *ends* of topological spaces and groups, and H. HOPF [1944] had proved an important theorem about the possible number of ends of spaces which also involves the definition of the ends of a countable group as an invariant of the group.

Finally, the theory of infinite algebraic extensions of a field, together with the emergence of more abstract definitions of a topological space, led, again in the 1940's, to the introduction of topological structurings of groups which are not based on any association of a cell complex with the group.

All of the topics mentioned here require a considerable technical apparatus for their full presentation. Also, most of their applications to combinatorial group theory appeared after the end of the period covered by our report. For these reasons, we shall confine ourselves here to a rather qualitative description of the ideas involved and we shall mention applications only if they admit a particularly simple formulation. However, we shall make an exception with the first application of cohomology theory which involves the theory of group extensions, and we shall begin our account with a brief outline of some aspects of this topic, although to do so is not entirely consistent with the mode of presenting the historical facts which we used in the earlier chapters.

Let Q and A be any groups. The problem of extension theory is the construction of all groups G which contain a normal subgroup isomorphic with A (and again denoted by A, since this will not lead to any confusion) such that the quotient group G/A is isomorphic with Q. One may recognize what probably is the first awareness of this problem in HÖLDER [1895] or DE SÉGUIER [1904, pp. 17–22]. But the first systematic approach is due to the Ph.D. thesis of O. SCHREIER [1926a and 1926b] which has already been mentioned in Chapter II.3. A concise and lucid account of the basic results may be found in Chapter III of ZASSENHAUS [1937 and 1949] or in KUROSH [1944], who observes that the information needed for the construction of all extensions G is of a very complex nature and becomes transparent only if one imposes additional conditions on A or on Q or on both and mentions, in this connection, the papers by SCHREIER [1926a and 1926b], BAER [1934], TURING [1938], and M. HALL [1938]. The latter paper is particularly important from the point of view of combinatorial group theory since it exhibits the role of the defining relators of Q (if Q is given by a presentation, i.e., as a quotient group of a free group) for the construction of the so-called *central extensions* G of A by Q. (See also Section 15.4 in M. HALL [1959]). In addition, M. HALL [1938] points out that one of the theorems needed in his paper deals with "a sort of algebraic-topological closure in semi-simple algebras." His use of the same type of concept in the theory of free groups and related groups (M. HALL [1950a]) will be discussed later in this chapter.

A particular question in the theory of group extensions which has found much attention arises from the concept of a *splitting extension*. We shall say that the extension G of a group A by a group Q *splits* if G contains a subgroup Q_0 isomorphic with Q such that the elements of Q_0 provide a complete system of coset representatives of A in G. In this case, G is also called the *semidirect product* of A and Q. (The direct product of A and Q is a very special case of a semidirect product.) In the theory of finite groups, splitting extensions have played a role for quite some time. For instance, if the orders of the finite groups A and Q are coprime, then G is always a splitting extension. (See, e.g., ZASSENHAUS [1937, p. 125]; the theorem is due to I. SCHUR.) For arbitrary groups, a fundamental theorem is due to ARTIN, whose proof was published by IYANAGA [1934] and in ZASSENHAUS [1937, p. 98]. Suppose that G is an extension of A by Q and that A is abelian. Then there exists a group G^* with a normal subgroup A^* such that

(i) $G^*/A^* \sim G/A = Q$
(ii) G^* is a splitting extension of A^* by Q
(iii) A^* is the direct product of A and another abelian group.

ZASSENHAUS mentions that this theorem can be generalized to cover the case of an arbitrary group A by using the theory of free products of groups. A generalization with explicit proofs was published by SEKI [1941]. The group G^* is called a *splitting group* (in German: *Zerfällungsgruppe*) of G. For the later literature, we have to refer to Sections 155 to 157 in G. BAUMSLAG [1974] and, specifically (for the theory of varieties), to P. Hall [1954b].

The formulas and definitions needed in extension theory simplify considerably if we assume that the group A is abelian. We shall do so from now on, and we shall write A additively, denoting its elements by a_i ($i = 1, 2, \ldots$). The elements of Q will be denoted by u_ν ($\nu = 1, 2, \ldots$). We shall now derive necessary conditions for the existence of an extension G of A by Q; these will then turn out to be sufficient conditions as well.

For every $u_\nu \in Q$ there must exist a coset $g_\nu A$ of A in G. Here g_ν is an element of G which is not uniquely determined since it can be replaced by any other element of the same coset. However, since A is abelian, the element

$$g_\nu^{-1} a_i g_\nu$$

depends only on the coset $g_\nu A$, i.e., on the element u_ν of the abstract group Q. We denote it by $a_i u_\nu$. The mapping

$$a_i \to a_i u_\nu$$

defines an automorphism of A, and the set of all of these automorphisms forms a group which is a homomorphic image of Q. We shall say that the mappings $a_i \to a_i u_\nu$ describe the *action of Q on A*; actually, A is now a $\mathbb{Z}Q$ module, where $\mathbb{Z}Q$ is the group ring of Q over the integers. If $a_i u_\nu = a_i$ for all ν and i, we shall say that Q *acts trivially* on A.

Suppose now that in Q,

$$u_\nu u_\mu = u_\lambda,$$

and consider the corresponding relation for g_ν, g_μ, g_λ. We will then have

$$g_\nu g_\mu = g_\lambda f(u_\nu, u_\mu),$$

where $f(u_\nu, u_\mu)$ is an element of A which depends on u_ν and u_μ. We may look upon f as a function of the ordered pair u_ν, u_μ of elements of Q. This function is called a *factor system* of Q in A; it cannot be chosen arbitrarily. The associative law for elements of G enforces the functional equation

$$f(u, v)w + f(uv, w) = f(u, vw) + f(v, w), \tag{$*$}$$

where now u, v, w are three arbitrary elements of Q. Every function f satisfying $(*)$ indeed defines an extension G of A by Q with the given action of Q on A. But it would be unreasonable to distinguish between extensions which differ only because we have chosen different representatives for the cosets of A in G. If two extensions can be changed into each other by replacing the g_ν by representatives $g_\nu' = g_\nu \cdot a_\nu$, where now a_ν are arbitrary elements of A and where $g_\nu \cdot a_\nu$ denotes the product of g_ν and a_ν in G, we call these extensions *equivalent*, and we use the same term for the corresponding factor systems. Now if $f'(u_\nu, u_\mu)$ is the factor system associated with the g_ν', we have

$$f'(u_\nu, u_\mu) = a_\mu + a_\nu u_\mu - a_\lambda + f(u_\nu, u_\mu)$$

(where $a_\nu u_\mu$ now denotes the result of the action of u_μ on a_ν). We can formulate this result as follows.

Let $\phi(u)$ be an arbitrary function of the elements u of Q with values $\phi(u)$ in A. Then the factor systems $f(u, v)$ and $f(u, v) - \phi(u)v - \phi(v) + \phi(uv)$ are equivalent.

Clearly, it must be true that the particular functions $f_0(u, v)$ defined by

$$f_0(u, v) = \phi(u)v + \phi(v) - \phi(uv) \qquad (\ast\ast)$$

with arbitrary $\phi(u)$ satisfy the functional equation $(\ast)$, and this can be verified by a simple calculation.

Since the functional equation $(\ast)$ is linear, it follows that the possible factor systems form a group under addition. The same is true for the particular factor systems defined by $(\ast\ast)$. The quotient group of these two groups will be denoted by

$$H^2(Q, A)$$

and is now called (for reasons to be explained below) the *second cohomology group* of Q with coefficients in A.

It must be noted that the order of $H^2(Q, A)$ does not reflect the number of nonisomorphic extensions G of A by Q with a given action of Q on A. The standard example (whose origin we have not been able to trace) is the following one. Let $p \geqslant 3$ be a prime number, let A be the direct sum of two cyclic groups of order p, and let Q be cyclic of order p. If G is nonabelian, we may always assume that the action of Q on A is given by the formulas

$$a_1 u = a_1 + a_2, \qquad a_2 u = a_2,$$

where u is a generator of Q and a_1, a_2 are suitably chosen generators of A. We have, of course,

$$pa_1 = pa_2 = 0, \qquad u^p = 1.$$

Now there exist exactly two nonisomorphic extensions of A by Q with this action of Q on A; in one of them, all elements of G have order p, and in the other, the preimage of u in G is of order p^2. However, the order of $H^2(Q, A)$ must be a positive power of p and, therefore, is greater than 2. The equivalence classes of extensions provide a more refined classification than the concept of isomorphism, but this refinement can be important even in problems of physics, as may be seen from the following example which we owe to a communication from L. AUSLANDER. The symmetry group of a crystal lattice is a finite extension of a free abelian group with three free generators. Such a lattice and its reflection in a plane always have isomorphic symmetry groups. But they need not be equivalent extensions, and crystals belonging to these different lattices may have different physical properties.

Before we introduce the concepts of homology and cohomology of groups, we have to say something about the origins of these concepts in topology. A good and, possibly, the earliest starting point for this purpose appears to be the famous paper by BERNHARD RIEMANN (1826–1866), on the theory of abelian functions, which appeared in 1857. This paper deals only with the simplest possible case, namely, surfaces (two-dimensional orientable spaces) and does not use the group concept at all. We quote one of his basic definitions:

> If it is possible to draw on a surface n closed curves which neither individually nor together form the complete boundary of a part of the surface but if they, together with any other closed curve, form the complete boundary of a part of the surface, then the surface will be called $(n + 1)$-fold connected.

This definition makes sense since the connectivity of a surface can be shown to be independent of the particular choice of the n curves. For a long time, the number n was called the first *Betti number* of the surface after ENRICO BETTI (1823–1892), who in 1871 introduced generalizations of RIEMANN's concept for higher-dimensional spaces and connectivities. Group theory enters homology theory in the paper by POINCARÉ [1895], who showed that the first Betti number is the rank of the maximal free abelian group which appears as a direct factor in the abelianized fundamental group π_1 of the space. This group, i.e., π_1/π_1', is now called the *first homology group* H_1 of the space. However, this statement is a theorem and not a definition since there exists a definition not only of H_1 but of all the homology groups H_n for arbitrary dimensions $n > 1$ which is independent of the fundamental (first homotopy) group. We shall not give the detailed definition but only the key terms for it (with the one-dimensional special case in parentheses). The first concept needed is that of a topological *simplex*, which, in one, two, and three dimensions, is, respectively, an interval (or edge), a triangle, and a tetrahedron. Next we need a simplicial decomposition of the topological space (in one dimension, this would be its representation as a graph). Then we need the additive free abelian groups S_n generated by symbols representing the n-dimensional simplices. Here each simplex has an orientation, and its negative is represented by the same simplex with the opposite orientation. Now we define *chains* of simplices (paths consisting of edges) and *closed chains* or *cycles* (closed oriented paths in one dimension). The cycles form a subgroup C_n of S_n. We define boundary operators δ_n which map a chain onto its boundary. Since boundaries are closed (i.e., $\delta_n\delta_{n+1} = 0$), δ_{n+1} maps all $(n + 1)$-dimensional chains onto cycles which form a subgroup B_n of C_n. The quotient group $C_n/B_n = H_n$ is then the *nth homology group* of the space. Of course, it has to be proved that it is independent of the simplicial decomposition of the space. HOPF [1942] used these concepts to define the second homology group $H_2(G)$ of any given group G by associating a simplicial complex with G. (HOPF calls $H_2(G)$ its *second Betti group*.) Later, HOPF [1945] defined the nth homology group $H_n(G)$ of any group G in a purely algebraic manner. The underlying idea of such a definition is sketched in the introduction to MACLANE [1963]. Definitions of homology groups of spaces with a minimum of technicalities and some historical remarks may be found in SEIFERT and THRELFALL [1934] and STILLWELL [1980, pp. 170–172].

The first application of the homology theory of groups was a surprising theorem proved by HOPF [1942]:

Let G be given by a presentation, i.e., as a quotient group F/R of a free group F. Then the second homology group $H_2(G)$ is given by the quotient group

$$(F'\cap R)/[F, R] = H_2(G),$$

where $F'\cap R$ denotes the intersection of the commutator group F' of F with R and $[F, R]$ denotes the normal subgroup of F generated by all commutators of an arbitrary element of F with an arbitrary element of R.

Of the applications of this theorem, we mention BAUMSLAG [1971, 1976a and 1976b], where the connection of the theorem with the problem of constructing finitely generated but infinitely related groups is investigated.

Leafing through Chapter 17 of BAUMSLAG [1974] will convince the reader that the literature on the cohomology of groups is considerably larger than that on the homology of groups. Although RIEMANN [1857] certainly did not even define the first cohomology group of a group (the group concept does not appear in his paper) and although we cannot provide any evidence which would show that his ideas have directly influenced later developments, an analysis of his paper provides a natural introduction to the subject.

Cohomology deals with the theory of functions on groups, and these functions must have certain properties which may be characterized as the properties of integrals taken over certain subspaces of a space associated with the group. In RIEMANN's paper, these integrals appear in the following context:

Let S be the Riemann surface of an algebraic function of a complex variable z. Topologically, S is equivalent to a closed orientable two-dimensional manifold. Assume that the genus g of this manifold is $\geqslant 1$. Then there exist g linearly independent algebraic functions $f_\nu(z)$ ($\nu = 1,\dots,g$) of z which are onevalued on S such that the line integrals

$$I_\nu(z, C) = \int_{z_0}^{z} f_\nu(\xi)\, d\xi, \qquad \text{taken over } C,$$

are finite for all values of z and taken over any arbitrary smooth curve C

which starts at a fixed point z_0 and ends at an arbitrary point z. The I_ν are called the abelian integrals of the first kind on S. They are not one valued on S, but their multivaluedness can be described completely by keeping z_0 fixed and choosing for C a closed curve C_0. The values of the resulting integrals $I_\nu(C_0)$ do not depend on C_0 itself, but only on its equivalence class in the homology group of S.

We shall not try to analyze the situation arising in RIEMANN's paper any further; our remarks are merely meant to introduce functions on a group which are not arbitrary but have the properties of integrals. Since integration is a linear functional, it follows that the values of integrals must be elements of an abelian group A which we will write additively. If Q is a group, let $x, y, z, x_1, x_2, x_3, \ldots$ denote arbitrary elements of Q. (In Q, we shall write group composition as multiplication.) We consider the elements of Q as points and any ordered pair (x, y) as an oriented interval. Similarly, a triplet (x, y, z) is considered a triangle with orientation, and so on. A one-dimensional integral $I_1(x, y)$ taken over the interval (x, y) is then a function of two "variables" x, y with values in A such that

$$I_1(x, x) = 0, \qquad I_1(x, y) + I_1(y, x) = 0.$$

Since integrals over a boundary are supposed to vanish, we have:

$$I_1(x, y) + I_1(y, z) + I_1(z, x) = 0.$$

So far, the group Q has not entered our definition except for the fact that Q is a collection of elements. We now assume that Q acts on A as a group of automorphisms. If α is any element of A, then there will be defined an isomorphic mapping of A onto itself by any element x of Q which we shall write as

$$\alpha \to \alpha x,$$

and we shall postulate that

$$I_1(x, y) z = I_1(xz, yz).$$

(Note that $I_1(x, y) \in A$.) This means that we consider the interval (xz, yz) as the image of the interval (x, y) under the action of z and that the integrals over these two intervals are related in a way which does not depend on the particular first interval (x, y), but only on the value of $I_1(x, y)$ and on the element z. In the case where the *action* of Q on A is *trivial*, i.e., if

$$\alpha x = \alpha \qquad \text{for all } \alpha \text{ and all } x,$$

the integral $I_1(x, y)$ is simply an invariant under the action of Q.

A function $I_1(x, y)$ satisfying all of the conditions mentioned is called a one-dimensional *cocycle*. Clearly, the cocycles form an abelian group under addition. They correspond to the one-dimensional integrals which vanish when taken over a boundary of a two-dimensional part of a manifold. Now it is trivially true that an integral has this property if its value, taken over an interval, is the difference of the values of another function taken at the endpoints of the interval. Correspondingly, we introduce a special type of functions $I_1^*(x, y)$ defined by

$$I_1^*(x, y) = J_0(y) - J_0(x),$$

where the values of J_0 are again in A and where

$$J_0(x) y = J_0(xy).$$

Such a function I_1^* is called a *coboundary*. It is easily verified that all coboundaries are cocycles, and, obviously, the coboundaries form (under addition) a subgroup of the group of cocycles. The quotient group of these two groups is called the *first cohomology group*, $H^1(Q, A)$. Its definition involves not only the groups G and A, but also the action of Q on A.

The second cohomology group of $H^2(Q, A)$ is defined in an analogous manner. We first define the *two-dimensional cocycles* $I_2(x_1, x_2, x_3)$ as functions of three elements of G with values in A such that

$$I_2(x_1, x_2, x_3) y = I_2(x_1 y, x_2 y, x_3 y),$$

$$I_2(x_1, x_2, x_3) = 0 \qquad \text{whenever } x_1 = x_2 \text{ or } x_1 = x_3 \text{ or } x_2 = x_3;$$

and

$$I_2(x_1, x_2, x_3) - I_2(x_1, x_2, x_4) + I_2(x_1, x_3, x_4) - I_2(x_2, x_3, x_4) = 0.$$

The last condition expresses the fact that a two-dimensional cocycle has the property of a two-dimensional integral which vanishes when taken over the surface of a tetrahedron with vertices x_1, x_2, x_3, x_4 and with consistent orientations of its faces. Again, in analysis it is true that a two-dimensional integral has this property if it can be expressed as an integral over the boundary of each face. In our context, this means that we can define particular two-dimensional cocycles I_2^* as follows: Set

$$I_2^*(x_1, x_2, x_3) = J_1(x_1, x_2) - J_1(x_1, x_3) + J_1(x_2, x_3),$$

where the values of J_1 are in A and

$$J_1(x_1, x_2) y = J_1(x_1 y, x_2 y).$$

The I_2^* are called *two-dimensional coboundaries*. The quotient group of the additive group of cocycles with respect to the additive group of coboundaries is called the *second cohomology group*, $H^2(Q, A)$, of Q with coefficients in A.

We can introduce new functions $f(x, y)$ of two group elements x, y of G which are again elements of A by setting

$$f(x_1, x_2) = I_2(x_1 x_2, x_2, 1),$$

$$I_2(x_1, x_2, x) = f(x_1 x_2^{-1}, x_2 x^{-1}) x.$$

The relations for I_2 stated above are then equivalent to the relations derived earlier for a factor system $f(x_1, x_2)$, provided that we normalize the factor system so that in the extension of A by Q, the coset representative of A is the unit element of Q. This results in the relations

$$f(x_1, 1) = f(1, x_2) = 0.$$

The function ϕ, which we had introduced in our account of extension theory, is connected with J_1 through the relations

$$x_1 = uv, \qquad x_2 = v,$$

$$J_1(x_1, 1) = \phi(uv), \qquad J_1(x_2, 1) = \phi(v),$$

$$J_1(x_1, x_2) = J_1(x_1 x_2^{-1}, 1) x_2 = \phi(u) v.$$

Of course, we now have $\phi(1) = 0$ since we must not change the representative of A in the extension of A by Q.

The interpretation of the theory of extensions of abelian groups in terms of cohomology theory was its first group-theoretical application. Many others were to follow later, and one of them will be mentioned below. However, our sketchy and elementary description of a few of the basic concepts of the theory must not give the impression that it developed slowly. The two fundamental papers by EILENBERG and MACLANE [1947] contain a wealth of theorems and introduce a very voluminous apparatus of concepts and formulas. In particular, the complete definition of all cohomology groups $H^n(Q, A)$ of any dimension n already appears in the first few pages of the first of these papers, together with the group-theoretical interpretation of $H^1(Q, A)$ and $H^2(Q, A)$.

The emergence of cohomology theory (not only of groups, but of various other algebraic or topological structures) as a new and highly abstract discipline had been prepared by two decades of a development both in algebra and in topology, which resulted in a strongly axiomatic and conceptual (i.e., calculation-avoiding) buildup of both fields. A careful analysis of this process might require a separate monograph. As for homology (including cohomology) theory, the book by MACLANE [1963] offers a good view of the new terminology and even the new type of formulas which accompanied its foundation and development.

Of the uses of cohomology theory in combinatorial group theory, we shall mention here only one paper by STALLINGS [1968] and refer for the literature (most of which was published after 1950) to Chapter 17, pp. 569–616, in BAUMSLAG [1974]. STALLINGS' paper also uses another topological concept, the *ends of a group*, which was introduced briefly in Chapter I.5. STALLINGS' results admit a remarkably simple formulation.

It is a near-trivial result that the extension of any group A (abelian or not) by a free group is always a splitting extension. From this, it follows that free groups have *cohomological dimension* 1, which means that all cohomology groups $H^n(F, A)$ of a free group F with respect to any abelian group A and with any action of F on A are trivial (i.e., $H^n(F, A) = 0$) for all $n > 1$. (It suffices to prove this fact for $n = 2$, in which case it is trivial to establish it for all $n > 2$.) Now STALLINGS showed that the free groups are the *only* finitely generated groups with this property. He also showed that a finite torsion-free extension of a finitely generated free group is again a free group and that a finitely generated torsion-free group which has infinitely many ends is a nontrivial free product. The last one of these theorems has already been mentioned in Chapter I.5. For generalizations of STALLINGS' theorems, see, again, BAUMSLAG [1974, pp. 605 − 606]. Concerning the concept of the ends of a group, we mention here, in addition to the remarks made in Chapter I.5, that it is a special case of a concept which applies to a much larger variety of structures (see FREUDENTHAL [1931 and 1945] and HOPF [1944]). Also, the definition of the number of the ends of a group which we gave in Chapter I.5, and which is based on the graph of a group arising from its definition through a presentation, cannot in any simple way be shown to be independent of the presentation which, in any case, must involve only a finite number of generators.

The last two topics to be discussed in this chapter are interrelated although of different origins. They involve the concepts of subgroup topologies and of the inverse limit of groups. The first one of these concepts emerged, in a sense, "naturally" from the increasingly abstract formulation of the basic concepts of topology and their transfer to the construction of algebraic systems. The concept of an inverse limit of groups appears naturally in the Galois theory of infinite algebraic extensions of fields.

That groups may be considered as topological spaces with the group elements as points of the space is, of course, an essential ingredient of the theory of Lie groups which were called continuous groups by LIE himself. The abstract formulation mentioned above consists of a definition of the neighborhood of a point which does not specify that such a neighborhood must share any properties with a solid ball in a finite-dimensional euclidean space. In particular, it may consist of only countably many points (elements of a countable group), and we can define a neighborhood of the unit element, for instance, as a normal subgroup of finite index in the whole group G, provided that the intersection of all of these normal subgroups is the unit element itself. The neighborhoods of an arbitrary group element g can then be defined as the cosets of these subgroups which contain g. Such a topology is called a *subgroup topology* of G.

Once a (reasonable) topology has been defined in a group G, we can *complete* the group under this topology, obtaining a group G^* which, in general, will have uncountably many elements even if G is only countably infinite. This process of completion of an algebraic structure was carried out first in the case of the field of rational numbers where the completion under a particular topology leads to the field of real numbers. That this is an application of the procedure outlined above becomes obvious if we use the method of WEIERSTRASS, who introduced real numbers by using sequences of nested intervals with rational endpoints where the length of the intervals tends to zero. (This method is, of course, equivalent to the more popular method of introducing the real numbers with the help of a Dedekind cut.)

KURT HENSEL (1861–1941) made the important discovery that every prime number p defines a particular topology for the field of rational numbers. The completions of this field were introduced by him as the fields of *p-adic numbers*. Certainly, neither WEIERSTRASS nor HENSEL used the terminology of topology. The development of the conceptual foundation of topology and its application to algebraic structures which, in the Nineteenth Century, nobody would have thought of as topological spaces cannot be described here. We can mention here only a few contributions to this development which have been selected mainly because they are of a systematic nature or have been influential. For completion theory, we cite VAN DANTZIG [1932], and specifically for topological groups, we mention PONTRJAGIN [1932, 1939]. The last reference is to a monograph which was cited by M. HALL [1950a], together with a large number of other relevant sources. HALL's paper represents the first systematic application of the concept of a subgroup topology to combinatorial group theory. It deals with groups G which have the property that, for any element $g \neq 1$ of G, there exists a subgroup of finite index in G which does not contain g, and which has a countable infinity of elements. This definition covers all countably generated free groups and it also happens to be identical with the definition of residually finite groups which was introduced by P. HALL as a special case of a residual property (see GRUENBERG [1957]).

M. HALL [1950a] also establishes the connection between subgroup topologies and the concept of a *projective* or *inverse* limit which is defined as follows:

Let G_s ($s = 1, 2, \dots$) be an infinite sequence of groups. (In general, s may run through all the elements of a directed set.) Assume that there exist homomorphic mappings

$$G_{s+1} \to G_s$$

for all s, and construct the group P whose elements are the sequences

$$(g_1, g_2, \ldots, g_s, g_{s+1}, \ldots),$$

where $g_s \in G_s$ and where, for all s, the homomorphic image of g_{s+1} is g_s. (The product of two such sequences is to be formed term by term.) If there exists a group G with normal subgroups N_s such that N_{s+1} is contained in N_s, $G_s \simeq G/N_s$ and the intersection of all N_s is the unit element, then P is the completion of G under the subgroup topology of G defined by the N_s.

(If we construct subgroup topologies where the neighborhoods of the unit element are normal subgroups of G, the postulate that these subgroups are of finite index in G guarantees that the intersection of any two of them is again of finite index.) For the literature on this subject, we refer to Section 9, pp. 375–380, of BAUMSLAG [1974]. We mention here only that sequences of groups G_s with the property described above appear automatically in the Galois theory of infinite algebraic extensions of a field. This topic had been investigated much earlier, e.g., by KRULL [1928], who also introduces a closure operator which has been recognized by HALANAY [1947] as the completion of the Galois group by means of a subgroup topology.

We already mentioned, in Chapter I.6, an important theorem due to TITS [1972] about the structure of finitely generated linear groups over fields of characteristic zero. The proof of this theorem uses the *Zariski topology* which provides an essential tool in algebraic geometry. We cannot go into details here, but we mention that, unlike the subgroup topology, the Zariski topology is based on a definition of *closed* sets and that it does not permit the separation of two points P_1 and P_2 by means of open sets S_1 and S_2 with empty intersection such that $P_1 \in S_1, P_2 \in S_2$. For information, see WEHRFRITZ [1973].

Chapter II.10

Notes on Special Groups

A large part of the literature on group theory deals with special groups; even if we confine ourselves to publications which deal with groups given by an explicit presentation in terms of generators and defining relators, the number of relevant papers is much too large to be considered in full in our historical account. Fortunately, there exists a monograph by COXETER and MOSER [1972] which provides as much information on this aspect of group theory as one may hope to find in a volume of reasonable size. Many of the groups listed in this monograph (which actually is the third edition of a book first published in 1957) are groups of finite order. However, the proofs that certain presentations define a finite group definitely belong in an account of the results of combinatorial group theory. Perhaps the most important example of results of this type is furnished by the investigation of the *groups generated by reflections* which are also called *Coxeter groups*. These are the groups with a finite number of generators R_i ($i = 1, 2, \ldots, n$), and defining relations

$$\left(R_i R_k \right)^{P_{ik}} = 1, \qquad p_{ik} = p_{ki};$$

$$p_{ii} = 1, \qquad i, k = 1, 2, \ldots, n.$$

COXETER and MOSER [1972] give a complete enumeration of the groups of this type which are directly indecomposable (i.e., not the direct product of subgroups of the same type) and are either finite or groups generated by reflection in a finite-dimensional euclidean space. These groups play an important role in the theory of finite simple groups and of finite-dimensional simple Lie groups and Lie algebras. See, e.g., BOURBAKI [1968]. Their investigation is mainly due to COXETER; for references, again see COXETER and MOSER [1972].

Next, we have to report on developments hinted at in Section I.6.D. They are based on two publications, by ARTIN [1925] and by NIELSEN [1927–

1931]. Both papers had long-range effects, but in very different ways, and they are of very different types and styles. ARTIN's paper is easy to read, at least for anyone who knows what a free group is. It introduces a new object into topology, and this is done in a charming intuitive way based on a waiving of the requirements imposed by the laws of full rigor. (The rigorous treatment was provided later by ARTIN [1947b].) It is a much-quoted paper, and it has been used many times in introductory or semipopular lectures on topology. The papers by NIELSEN [1927–1931], although carefully and clearly written, are nevertheless not easily accessible. In part, this is due to their length—a total of 327 pages. Apart from this, proofs of several basic theorems require a perfect knowledge of non-euclidean geometry, something which is not so easily acquired. Isolated theorems proved in NIELSEN's papers have been widely quoted, but the strongest evidence for their long-range importance is furnished by the fact that now, 50 years after their appearance, an English translation is being prepared for publication.

Although at the time it was published nobody would have guessed that ARTIN's paper is also relevant to one of the topics of NIELSEN's papers, namely, the theory of the groups of mapping classes of two-dimensional manifolds. This fact has now been well established and is documented in the title of a monograph by BIRMAN [1975]. For the definition of topological concepts, including those of a braid or a mapping class, we have to refer to this monograph, confining our own account to some of the group-theoretical aspects of the papers by ARTIN and NIELSEN.

Although a braid is a topological object, the group B_n of *n-string braids* can be defined in a purely algebraic manner as the group of automorphisms β of a free group F_n on free generators x_ν ($\nu = 1, \ldots, n$), which have the property that $\beta(x_\nu)$, the image of x_ν under the action of β, is a conjugate of one of the generators x_μ, where the mapping $\nu \to \mu$ defined a permutation of the symbols $1, 2, \ldots, n$, and where

$$\beta(x_1 x_2 \cdots x_n) = x_1 x_2 \cdots x_n.$$

Clearly, B_n is a subgroup of B_{n+1} for all n. Also, B_n contains a normal subgroup B_n^* of index $n!$, called the *unpermuted braid group*, whose elements are defined by the fact that each x_ν is mapped onto a conjugate of itself. ARTIN derived (by topological methods) a presentation of B_n in terms of generators and defining relations and showed that presentations for the groups of all knots or linkages can be obtained by choosing a suitable n and a suitable β in B_n and considering the groups with generators x_ν and defining relations

$$\beta(x_\nu) = x_\nu \qquad (\nu = 1, \ldots, n).$$

On the other hand, every group with such a presentation is the group of a knot or a linkage, a projection of which can be read off the projection of β

(as a topological braid) onto the euclidean plane. The next paper on braid groups is due to BURAU [1933], who derived (from ARTIN's presentation of B_n) a presentation of B_n^* and later also found (BURAU [1935]) an $(n-1)$-dimensional matrix representation for B_n with entries from a ring of polynomials with integral coefficients in an indeterminate x and its inverse x^{-1}. (The question of its faithfulness for $n > 3$ is still open.) Another early group-theoretical paper on braid groups is the one by MAGNUS [1934c] which contains both a derivation of ARTIN's presentation of B_n from its algebraic rather than its geometric definition (a derivation which was rediscovered by BOHNENBLUST [1947]) and a presentation of the mapping class group of the sphere with n boundary points as a quotient group of B_n with a geometric interpretation of the connection between the two groups. The next group-theoretical investigation seems to be a paper by MARKOFF [1945] which contains many results, in particular, a group-theoretical analysis of B_n^* which turns out to have a descending normal series of length n in which the quotient groups of consecutive groups are free of ranks $1, 2, \ldots, n - 1$. The same result also appears in the paper by ARTIN [1947a] which has been mentioned already. We cannot go into the details of these and other papers (e.g., ARTIN [1947b]) which appeared at about the same time or later and refer to Section 256 of BAUMSLAG [1974] and, especially for the group-theoretical aspects, to the survey by MAGNUS [1973] which, at the end, also mentions the interpretation of the braid groups as special cases of a new class of groups associated with the Coxeter groups. For the generalization of the braid groups which connects them with the general theory of mapping class groups of two-dimensional manifolds, we have already mentioned the monograph by BIRMAN [1975] which stresses the topological aspects.

The last reference brings us back to the papers by NIELSEN [1927–1931]. His papers deal with the topological self-mappings of closed orientable two-dimensional manifolds of a genus $g > 1$. The fundamental group Φ_g of such a manifold has a presentation with $2g$ generators $a_i, b_i, i = 1, 2, \ldots, g$, and a single defining relator

$$a_1 b_1 a_1^{-1} b_1^{-1} a_2 b_2 a_2^{-1} b_2^{-1} \cdots a_g b_g a_g^{-1} b_g^{-1}.$$

NIELSEN showed that the mapping class group M_g of such a manifold is isomorphic with the group of outer automorphisms (i.e., the quotient group of the automorphism group with respect to the group of inner automorphisms) and proved that all automorphisms of Φ_g can be induced by those automorphisms of the group freely generated by a_i, b_i which map the relator onto a conjugate of itself or of its inverse. He also extensively investigated the outer automorphisms of finite order in connection with the *fixed-point problem* of the mappings. This is, in a way, a continuation of NIELSEN's Ph.D. thesis which we mentioned in Section I.6.D. However, the transition from $g = 1$ to $g > 1$ poses tremendous difficulties. It is even difficult to find generators for M_g. This was, in complete generality, done first by DEHN [1938], whose results were rediscovered and deepened by LICKORISH [1964]. Both of these authors use topological methods. A characteristic of much of NIELSEN's approach is the use of non-euclidean geometry. The group $\bar{\phi}_g$ is represented in the classical POINCARÉ–KLEIN–FRICKE tradition as a discontinuous group of Moebius transformations which map the unit disk of the complex plane onto itself. The points on the boundary of the disk are then limit points of the images of a given point in the interior under the action of the elements of Φ_g. NIELSEN introduces specific infinite sequences of elements of Φ_g to define particular points on the boundary of the disk. One may consider a similar procedure introduced by FRICKE (in FRICKE and KLEIN [1897, pp. 415–428]; for an English version, see MAGNUS [1974a, Chapter 4] as a forerunner of this artifice. However, NIELSEN's arguments exceed FRICKE's by an order of magnitude in sophistication and difficulty.

The current interest in NIELSEN's work is being maintained in part by its importance for the comprehensive theory of kleinean groups which has been developed in the past decades and which is completely outside the range of our account. As a reference, we mention BERS and KRA [1974]. We also note that the theory of homomorphic mappings of Φ_g is important for the construction of 3-manifolds. As an example, we mention that the question raised by POINCARÉ mentioned in Chapter I.4 (the so-called *Poincaré conjecture*) can now be formulated in purely group-theoretical terms as a problem involving the homomorphic mappings of Φ_g onto the direct product of two free groups of rank g. For a formulation of this result and for references, see LYNDON and SCHUPP [1977, p. 195].

Studies of the special groups which motivated the development of combinatorial group theory never stopped, although there are fertile as well as barren years for their investigation. For instance, there are many contributions to the theory of fuchsian groups for which the groups Φ_g mentioned above are the most prominent examples. For these, we cite only three papers by HOARE, KARRAS, and SOLITAR [1971 and 1973] because they mark an instance where group theory is, finally, catching up with topology. From the classification of closed two-dimensional orientable manifolds of finite genus and from the identification of the subgroups of fundamental groups of a manifold with the fundamental groups of its unramified covering spaces, it follows immediately that every subgroup of finite index of a group Φ_g is a group Φ_k with $k \geqslant g$. This and many other results for fuchsian groups were proved by HOARE, KARRASS, and SOLITAR in a purely group-theoretical manner. Their method of proof is also of interest. They introduced generalizations of a free product with amalgamations which they called *tree products*, a concept which anticipates some of the work in SERRE [1977]. Also, their papers confirm an old contention of DEHN that there exists a topological component in the intrinsic structure of the fundamental groups themselves.

The study of the groups of knots and linkages as well as that of fundamental groups of three-dimensional manifolds (which began in earnest with the book by SEIFERT and THRELFALL [1934]) also has continued. The papers by BRAUNER [1928] and BURAU [1932, 1934, and 1935] are examples although, by far, not all of them. Apart from the monographs by REIDEMEISTER [1932a] and CROWELL and FOX [1963], we also mention here the one by NEUWIRTH [1965]. Important results have been obtained with topological methods by HAKEN [1962], WALDHAUSEN [1968] and others. For the very different situations arising in 4, 5, and higher dimensions, see also HAKEN [1973] and KERVAIRE [1965].

Additional historical information about the interaction between topology and combinatorial group theory may be found in the monograph by STILLWELL [1980].

With these rather vague hints, we leave the theory of special groups arising from topology and turn to the special infinite groups which have been investigated by using a transfer of methods developed in the theory of finite groups. Again, we have to restrict ourselves to giving a few hints. Probably the most easily formulated and also rather early results were those obtained by HIRSCH [1938 to 1954], who introduced a class of infinite groups which he called *S-groups* and which now are known as *polycyclic groups*. They are generalizations of the finite solvable groups in the sense that they contain a finite normal series with cyclic quotient groups. If the whole group is infinite, there must occur infinite cyclic groups among these quotient groups. Although the Jordan–Hölder theorem will not hold, the number of these infinite cyclic quotient groups is an invariant of the whole group. And a combination of the papers by MAL'CEV [1949] and AUSLANDER [1967] proves that the polycyclic groups are exactly the solvable subgroups of the finite-dimensional matrix groups with entries from the ring of integers.

In Chapter II.4 we mentioned KUROSH's idea of introducing *local properties* of groups, i.e., properties which are shared by all finitely generated subgroups but not necessarily by the whole group. This has been an eminently fruitful way of defining classes of special groups which are general enough to be of interest and yet accessible to a structural analysis. For instance, KUROSH [1939] introduced locally free groups and a year later FOUXE-RABINOVITCH [1940] proved that these groups cannot be simple. But the theory even of this local property has sprouted in recent years, and the theory of locally finite groups (which contains the Burnside groups as a very special case) is voluminous. Also, in Chapter II.4 we mentioned the many results obtained by R. BAER in the theory of infinite groups which we cannot include in our account. The three papers by BAER [1945] exemplify our reasons for this omission. Here BAER studies the possibilities of representing a given group as a quotient group. His starting point is the remark that every group is a quotient group of a free group, and BAER succeeds in introducing some effective classification into his topic. But his papers cover 125 pages and a very large number of theorems. Even his shorter paper, BAER [1949], on groups with descending chain condition for normal subgroups has many results which cannot be formulated briefly and is succeeded by numerous related papers by himself and others which are listed in Section 89 of BAUMSLAG [1974], whose classified survey of the literature from 1940 to 1970 is the only adequate source of information available for these and related topics.

We conclude this chapter with a few remarks about the *Heisenberg group*. Actually, this name covers a class of groups in which the commutator subgroup belongs to the center. We have not located the origin of this term. What we wish to note is the fact that groups of such a simple type play a large role in a surprisingly large number of mathematical disciplines. A survey by HOWE [1980] lists these but deals only with the application of the Heisenberg group in functional analysis. Other important applications appear in a monograph by AUSLANDER and TOLMIERI [1975].

Chapter II.11

Postscript: The Impact of Mathematical Logic

In all of the previous chapters we have dealt with developments which started before the middle of the Twentieth Century, although, in many cases, we also reported on some of their later phases. In the present chapter we are making an exception to this rule which we imposed on our account in order to keep it from becoming too voluminous. Our reason for doing so is the exceptional nature of the impact of mathematical logic on combinatorial group theory. We encounter here the situation where concepts and methods from one mathematical discipline become essential ingredients of the theorems and of the answers to previously open questions of another discipline. Of course, events of this type have happened before, but they deserve to be mentioned with distinction. In a way, the connection between topology and combinatorial group theory provides a similar example. However, there exists a notable difference between these two occurrences. Topology and combinatorial group theory grew up together, but the event we are now going to describe involves the action of a highly developed sophisticated theory on another equally developed discipline.

In Chapter I.7 we briefly mentioned DEHN's emphasis on decision problems and the fact that the same type of question had been raised by HILBERT in his Tenth Problem which seeks a method of deciding whether any given diophantine equation has a solution or not. The answer to a decision problem is an *algorithm* or a *general and effective procedure*. All one could say about these concepts at the time of HILBERT and DEHN was that one knew one when one saw one. It is one of the great achievements of Twentieth Century mathematics to have given a precise meaning to these concepts and thereby make it possible to prove that there exist *unsolvable decision problems*. We shall not give a definition of these terms and we shall also abstain from defining the only technical terms of the theory which we shall need below in order to formulate HIGMAN's theorem, namely, the terms *recursively enumerable* and *recursive* which can be applied to sets of freely reduced words in finitely generated free groups. Instead, we refer the reader to pp. 298–321 in ROTMAN [1973] whose account of the theory is geared specifically to the needs of group theorists. So is the partly historical paper by STILLWELL [1982] which deals specifically with the word problem and the isomorphism problem. The history of the theory of recursive functions has been outlined by KLEENE [1981]. Both STILLWELL and KLEENE cite the literature and describe the fundamental contributions of CHURCH, GÖDEL, POST, TURING, and others.

We now start with a result which is known as the

Novikoff–Boone Theorem. *There exists a finitely presented group with an unsolvable word problem.*

The full, detailed proof by NOVIKOFF appeared in 1955. However, he had announced his result in a short note without proof three years earlier. The proof by BOONE is contained in a sequence of papers, BOONE [1955]. It should perhaps be mentioned that NOVIKOFF was born in 1901 and, although he was a mathematician of high repute, he had not published anything of this importance before. His later contribution to the Burnside problem has been mentioned briefly in Section I.6.F. MAX DEHN, the discoverer of the word problem, had died shortly before the publication of the papers by NOVIKOFF and BOONE in 1952 at the age of 72.

The discovery of finitely presented groups with an unsolvable word problem was followed by a large number of proofs for the general unsolvability of other group-theoretical problems. A monograph by CHARLES MILLER III [1971] provides a survey of the results obtained up to that time. We mention here only a later unpublished result of the same author, which exhibits the great complexity of problems of this type. There exists a finitely presented group G with solvable word and conjugacy problems which contains a subgroup H of index 2 with an unsolvable conjugacy problem. (Of course, the word problem for H must also be solvable.) On the other hand, it is possible that the conjugacy problem is solvable for a finitely presented group H but not for a group G containing an isomorphic replica of H as a subgroup of index 2. (Of course, G also has a finite presentation and a solvable word problem.)

What we have described so far is a new and very important aspect which has been added to the problems of combinatorial group theory through the discoveries of mathematical logic. But this is only the beginning of the story. To explain what we mean, we shall go back to the beginning of combinatorial group theory.

The papers by DYCK, DEHN, TIETZE, and others deal almost exclusively with finitely presented groups. The fact that such groups may contain infinitely generated subgroups was first noted by DEHN [1911], and the theory of one-relator groups immediately produces subgroups of some of these groups which are both infinitely generated and infinitely related. The paper by B. H. NEUMANN [1937b] shows that there exist uncountably many nonisomorphic two-generator groups which are infinitely related. All of this seems to indicate that being finitely presented (or presentable) is a rather inconsequential property of a group. And yet this is not true if we use the language of mathematical logic. This insight is contained in a theorem due to G. HIGMAN [1961] which states:

A finitely generated infinitely related group G can be embedded as a subgroup in a finitely presented group if and only if the set of relators of G (as a set of freely reduced words in the generators) is recursively enumerable.

We note here that in HIGMAN's proof, the relators must be words which involve only positive powers of the generators. But this can always be arranged by using a trick introduced by DYCK [1882]: We adjoin one additional generator and one additional relator which is the product of all of the generators. Then the inverse of each generator can be replaced by a product of the remaining generators.

HIGMAN's theorem characterizes the finitely generated subgroups of a finitely presented group. There is no way known of doing this without using the terminology of mathematical logic. In addition, the proof of HIGMAN's theorem provides some surprising examples of finitely presented groups. We mention two of them:

There exists a finitely presented group which contains isomorphic replicas of all countable abelian groups as subgroups.

There exists a finitely presented group which contains isomorphic replicas of all finitely presented groups as subgroups.

However, there cannot exist a countably generated group which contains isomorphic replicas of all finitely generated groups as subgroups. This follows from the result of B. H. NEUMANN [1937b], according to which there exist uncountably many nonisomorphic two-generator groups, whereas a countably generated group contains only countably many two-generator subgroups. This is an application of GEORG CANTOR's theory of transfinite numbers which is, in a sense, more interesting than its original application which provided a simple proof for the fact that there exist transcendental numbers. After all, LIOUVILLE had earlier constructed an explicitly given set of uncountably many transcendental numbers. But so far there is no way of proving the nonexistence of the finitely generated group mentioned without using CANTOR's theory.

The next topic in combinatorial group theory into which the concepts of mathematical logic enter in an essential way is that of algebraically closed groups. The starting point is a paper by B. H. NEUMANN [1943a] which deals with the problem of solving equations in groups. This is the analog of the problem of solving algebraic equations with coefficients in a field. But the formulation of the problem is more complex. BAUMSLAG [1976, pp. 442–445] lists reviews of 16 papers dealing with this problem until 1970. Of these, we mention here only GERSTENHABER and ROTHAUS [1962] because it applies the theory of compact connected Lie groups rather than methods of combinatorial group theory to the problem. After it had been stated, it was possible to transfer the concept of an algebraically closed field to the concept of an algebraically closed group. This was done by W. R. SCOTT [1951]. We follow LYNDON and SCHUPP [1976, pp. 227–228], introducing the necessary definitions:

Let G be a group and let g_k ($k=1,2,\ldots$) be elements of G. Let x_j ($j=1,2,\ldots$) be "variables" or "indeterminates," and let $W_i(x_j, g_k)$ ($i=1,\ldots,m$) and $V_l(x_j, g_k)$ ($l=1,\ldots,n$) be finitely many words in the x_j and g_k. Then a system of equations and "inequalities"

$$W_i(x_j, g_k) = 1, \qquad V_l(x_j, g_k) \neq 1 \tag{$*$}$$

is said to be *consistent with* G if there is a group H and an embedding ϕ: $G \to H$ of G into H such that the system ($*$) has a solution in H (where now, of course, the x_j are elements of H).

A group A is called *algebraically closed* if every finite set of equations and inequalities which is consistent with A already has a solution in A.

The simplest (and very typical) example of an equation which is not consistent with a given group G is the following. Suppose a, b are elements of finite but different orders in G. Then the equation

$$x^{-1}axb^{-1} = 1$$

is not consistent with G since it implies that a and b must have the same orders.

SCOTT [1951] proved that every countable group can be embedded as a subgroup in an algebraically closed group which is again countable. He also proved that all finite groups have isomorphic replicas as subgroups in any countable algebraically closed group. On the other hand, there must exist uncountably many countable algebraically closed nonisomorphic groups according to B. H. NEUMANN [1937b] (see above). The question arises: Which groups will appear as subgroups in all countable algebraically closed groups? For finitely generated groups, this question is answered by the following theorem.

A finitely generated group G can be embedded in every countable algebraically closed group if and only if G has a solvable word problem.

The "if" part of this theorem is due to B. H. NEUMANN [1973], and the "only if" part is due to MACINTYRE [1972]. For more information about the theory of algebraically closed groups, see B. H. NEUMANN [1973] and LYNDON and SCHUPP [1976, pp. 227–234].

It has been noted by B. H. NEUMANN [1973] that the theorem stated above may imply that although we know of the existence of uncountably many nonisomorphic countable algebraically closed groups, it may be impossible to construct explicitly at least two such groups which are not isomorphic.

The last item to be described in this chapter involves the role of finitely presented infinite simple groups. There is no lack of known countably infinite simple groups; they can be constructed easily by using groups of invertible $n \times n$ matrices with $n \geqslant 3$ and entries from a countably infinite field. See, e.g., ROTMAN [1973, pp. 165–173] for proofs. (The results are much older; we did not find out how old.) But the existence of a finitely generated infinite simple group already posed a difficult problem which was first solved by HIGMAN [1951]. Shortly afterward, CAMM [1953] constructed an uncountable number of two-generator nonisomorphic simple groups. But the problem of finding a finitely presented infinite group was first solved by R. J. THOMPSON in 1969. His paper has not been published, but shortly afterward, HIGMAN [1973] explicitly exhibited a (countable, of course) infinity of nonisomorphic finitely presented infinite groups. All of these results would properly have belonged in the previous chapter on special groups were it not for the fact that these groups are connected with groups which have solvable word problems. We quote the latest result from a paper by R. J. THOMPSON [1980] which also contains references to the earlier literature:

> A finitely generated group has a solvable word problem if and only if there is an embedding of the group (such that two elements not conjugate in the original group will not be conjugate in the embedding group) into a finitely generated simple group which is a subgroup of a finitely presented group.

Hilbert's Tenth Problem, formulated in 1900 at the International Congress of Mathematicians in Paris, asks for an algorithm to decide whether any given diophantine equation has a solution. (A diophantine equation is an equation of the form

$$P(x_1,\ldots,x_n) = 0,$$

where P is a polynomial in n variables $x_1,\ldots,x_n$ with integral coefficients, and a solution consists of a set of n integers which satisfy this equation.) It took the greater part of the Twentieth Century and the efforts of many prominent mathematicians to show that such an algorithm does not exist. The crowning achievement of these efforts is the paper by MATIYASEVICH

[1971] which gives the answer in a surprising form. It is remarkable that the solution of HILBERT's number-theoretical problem can also be used to prove that there exist finitely presented groups with an unsolvable word problem. Such a proof, with references to the literature, may be found in LYNDON and SCHUPP [1976, pp. 217–227].

Although our long historical account deals only with a small part of mathematics, and although it is very far from being complete, it exhibits a great proliferation of problems and lines of research. The discovery that the answers to HILBERT's problem in number theory and the word problem in group theory share a common ground confirms our hope that, nevertheless, something like the unity of mathematics will continue to exist.

Chapter II.12

Modes of Communication

This chapter is a continuation of a part of Chapter I.8. We shall try to describe the development of the modes of communication since the end of World War I in 1919.

The year 1919 was chosen because it marks the emergence of a new era for combinatorial group theory, something which does not apply to most of the other disciplines of mathematics. Other important changes occur only after World War II.

From the middle of the Nineteenth Century to the present, the principal instrument for the documentation of mathematical research has been mathematical and related journals. Their number increased from about 180 in 1920 to about 600 in 1948 and to about 1600 in 1980. These numbers have been taken from the listings in the *Jahrbuch für die Fortschritte der Mathematik* and the *Mathematical Reviews*, but they do not completely reflect the actual growth of the volume of mathematical research. For the period from 1940 to 1980, the table at the beginning of Chapter II.14 is a much better guide. It shows that in these years the number of published papers increased twelvefold, i.e., much faster than the number of journals. The reason for this is twofold. The journals (or, at least, many of them) became more voluminous. Simultaneously, the number of shorter papers increased.

The rising costs of mathematical journals (even of those supported by page charges to the authors and published by mathematical societies) drastically reduced, over a period of a few decades, the number of private subscribers to journals. This had the effect that today's mathematician (with the unlikely exception of a very wealthy one) depends for his or her work very heavily on access to a university library or to an outstanding public library. Also, it is only the extraordinarily rapid increase in the number of university libraries in many countries which has kept most journals viable, particularly those produced by commercial publishers.

As indicated above, all of the developments described so far took place only after the end of World War II in a really decisive form. Their effects were mitigated by improvements in the methods of copying of printed or handwritten material. In particular, the invention of xerography made the acquisition of mathematical documents incredibly easy and inexpensive and also simplified their storage since one could obtain copies of a small part of a book or journal without having to obtain a whole volume. This resulted in the emergence of a new form of limited publication through so-called preprints which could be made from a typewritten text and sent to a not too large number of fellow mathematicians.

The role of books, monographs, and reviewing journals for the period under consideration will be discussed (in a slightly different context) in Chapter II.14. Letters, which we also mentioned briefly in Chapter I.8, certainly continued to be used for exchange of information and for inquiries, but their use for the purposes of exchanging ideas or maintaining scientific contacts probably declined after World War II. Whereas this statement is not more than a guess based on the personal experiences of one of the authors, we can document the increasing importance of the most elusive of oral communications, namely, the informal discussions between active postdoctoral research mathematicians. The little statistical table at the beginning of Chapter II.14 shows that the percentage of jointly authored papers has about quadrupled in the four decades from 1940 to 1980 and that the absolute number of such papers was, in 1980, more than twice the total number of publications in 1940. We know, of course, that not all joint papers are based on oral discussions. Correspondence and also the discovery that two people had done the same work independently and at the same time may also lead to a publication with several authors. Nevertheless, we maintain that the rapid increase in the number of joint papers is mainly due to informal discussions. In support of this statement, we quote an article by E. C. ZEEMAN which appeared in 1967 in the May issue of the *New Scientist*. There the author explains the reasons for the foundation of the Mathematics Research Centre at the University of Warwick at Coventry, England. ZEEMAN is the founder and director of this Institute which, since 1966, has attracted a large number of mathematicians to its programs whose topics vary from year to year. In his article (and, in greater detail, in some unpublished memoranda) ZEEMAN explains the need for informal conversations between mathematicians and his proposals to provide opportunities for this type of communication. In his memoranda, ZEEMAN also mentions the importance of oral communication in two other research institutes: The Institute for Advanced Study in Princeton, NJ and the Institut des Hautes Etudes near Paris. The Princeton Institute was founded in 1930 by two private sponsors who were much impressed by a memorandum of ABRAHAM

FLEXNER who subsequently became its first director. FLEXNER's thesis was that the high teaching load at American universities hampered the full development of the sciences in the United States, pointing out the much more favorable conditions for research work at European universities. Indeed, OSWALD VEBLEN, the first head of the mathematics department of the Princeton Institute told MAGNUS that, at the highly prestigious Princeton University, he had taught 18 hours a week and had welcomed the opportunity of finding more time for research offered to him at the Research Institute. For a variety of reasons (the existence of the Institute for Advanced Study probably being one of them), the teaching load for research mathematicians has dropped drastically since World War II, and now at least, one of the main functions of the Institute is indeed the one outlined by ZEEMAN.

The Institut des Hautes Etudes near Paris was founded in 1958 with the support of the industry of the European Common Market. It functions in a manner similar to that of the Princeton Institute, offering its facilities to a large number of mathematicians from many countries.

Research institutes for mathematics are definitely a novelty in the sense that they appeared only after World War I, although they existed for other sciences well before that time. All of them facilitate personal contact between mathematicians, and we mention here two others which cultivate international relations. They are the Tata Institute of Fundamental Research at Bombay, India and the Forchungsinstitut fuer Mathematik at Oberwolfach in the Federal Republic of Germany. The Tata Institute was founded in 1945 by members of a wealthy family of industrialists. It has a smaller number of invited visitors than the institutes mentioned above, with a preference for distinguished scholars who also give formal lectures during their stay. The Institute at Oberwolfach specializes in the arrangement of conferences of one-week duration with a definite program of lectures delivered by the participants. It was founded in 1943 by WILHELM SUESS but became effective only after 1950 through the financial support of German industry. For many years, its conferences on group theory were strongly influenced by REINHOLD BAER whose name will appear prominently in the next chapter.

Like research institutes, specialized conferences are a novelty in mathematics, appearing in rapidly increasing numbers and in many countries only after World War II and replacing, in effectiveness, not only the international congresses of mathematicians but, to some extent, even the meetings of the various mathematical societies as a means of communication. Of course, they owe their usefulness and even their existence to the proliferation and specialization of mathematical research. The same is true for specialized journals which began to appear (with a few exceptions) only somewhat later. They usually cover wider fields of research than some conferences. (There exists a Journal of Algebra but, so far, not a Journal of Group Theory.)

Apart from conferences and organized meetings of various types, other opportunities for personal contacts between mathematicians were created after World War II by an abundance of visiting professorships, fellowships, and exchanges of scholars. These were financed by government agencies, private foundations, and universities. Some of these institutions also financed travel expenses for the purpose of visiting conferences. (In the 1950's, these opportunities were offered mainly through American institutions.) The emergence of air travel—very rare before World War II—made even transoceanic journeys feasible for comparatively short visits. Also, the cost of travel relative to the income of members of the academic profession had decreased considerably.

As a last item in our long list of changes which facilitated oral communications, we have to mention the fact that not only did new universities appear in many countries, including those with a long history of academic research, but that the size of mathematics departments also increased considerably in many places. It now became possible for many departments to employ a group of mathematicians working in the same specialized field. This situation existed even before 1940 in a few outstanding universities, but in the decades after 1950, it became more common. Here we also have to mention the Steklov Institute in Moscow. It is a research institute without an extensive program for visitors and of a type rather common in other, particularly, in experimental sciences, where large laboratories form the basis for a close cooperation of scholars. On a purely academic basis, such research institutes have existed in many fields since the turn of the century (in Germany, under the name of KAISER WILHELM—now Max Planck-Institutes), but not in mathematics. Several mathematicians mentioned prominently in our report, e.g., ADIAN, KOSTRIKIN, NOVIKOV, ŠAFAREVIČ, were or are members of the Steklov Institute.

In Chapter I.8, we mentioned the effects of the student-teacher relationship existing in universities but not in academies. Throughout our book on various occasions, we have mentioned the Ph.D. advisers and the fields of the thesis of mathematicians who have played an important role in combinatorial group theory. All of these mathematicians had become active in research before 1935. We have tried to update this information by asking approximately 90 mathematicians of later generations how and when they were introduced to mathematics in general and to our special field in particular. We have also looked at their later publications to find out how lasting the influence of the Ph.D. adviser was. It would be difficult to use the data we obtained as a basis for statistical tables. Reporting instead on our general impressions, we might say that there are few if any surprises.

The people who became interested in mathematics not in high school, but only after entering a university form a small but notable minority. The interest in a specialty has been aroused frequently by personal contact with a senior mathematician, the choice of a thesis adviser being an obvious example. But it could also be induced by reading lecture notes written and distributed on the other side of the Atlantic. The influence of the Ph.D. thesis on later research work has increased. This is simply an expression of the increasing tendency towards specialization which, in turn, is largely due to the fact that the efforts needed for acquiring the basic knowledge of a field have grown rapidly during the past 50 years. Nevertheless, the interaction between different branches of mathematics has not stopped, and we have made an effort to document this fact in our book.

Although mathematical publications are using ideograms like $\exists, \forall, \in$, etc. to an increasing degree, all communications in mathematics are still based on the use of one of the many natural languages. In Chapter I.8, we sketched the linguistic situation up to World War I. In the following years, things changed slowly but not drastically. English began to replace French as the most widely understood language, but English, French, German, and Italian, the four official languages of international congresses, were still sufficient to cover most of the literature, in spite of the fact that authors with many other native languages, e.g., Polish, Japanese, Hungarian, etc., began to appear in rapidly increasing numbers. In particular, Russian-speaking authors continued to publish at least their most important papers in one of these languages until about 1938 when they rather abruptly began to use Russian exclusively. For a while the effects of this change were hidden by the reduction of most international communication during and shortly after World War II. During the next two decades, English and Russian became the dominant languages in mathematical literature. This may be true to a greater or lesser extent for different fields; in combinatorial group theory, the phenomenon is perhaps particularly strong, covering about nine-tenths of all papers. For a variety of reasons, the number of English speaking mathematicians who have a reading knowledge of Russian has always been small (although it has been slowly increasing). This would have led to very serious difficulties without the translation project started by the American Mathematical Society in 1949 which ever since has made an important number of Russian research papers accessible to English-reading mathematicians. Over the years, the time lag between the publication of the Russian original and the English translation has diminished considerably. Apparently, a reading knowledge (if not more) of English is very common among authors of Russian research papers.

The emergence of English as an international language of mathematics is particularly noteworthy in view of the fact that after World War I and again after World War II, an increasing number of countries developed institutions of higher learning which produced mathematical research. We are not prepared to analyze in detail the causes of the preponderence of English. The most obvious one has been mentioned already. During two decades after World War II, American institutions extended and financed invitations for visits to the United States to a large number of mathematicians from many countries.

Our description of the linguistic situation in mathematics is, as we believe, essentially correct with respect to combinatorial group theory. For mathematics in general, the situation is probably more complex and less comforting. For instance, in the years 1960–1967, there appeared a Chinese journal (written, of course, in Chinese characters) which the American Mathematical Society deemed important enough to be translated in its entirety. The translation appeared in nine volumes under the title *Acta Mathematica Sinica*. Except for Japanese mathematicians, the vast majority of non-Chinese mathematicians must have found the original incomprehensible.

So far, we have described only the problems arising from the language of research papers. Text books (even rather elementary ones) and monographs are being translated from many languages into many others, and this includes numerous translations from English into Russian (and, of course, vice versa). In many countries, the student of mathematics can acquire a knowledge of the fundamental results of practically all disciplines through text books written in his or her native language.

Probably the greatest linguistic obstacle impeding communication between mathematicians is the language of mathematics itself. The coining of new technical terms is an absolute necessity for the evolution of mathematics. The rather straightforward looking statement: "There exist finitely presented infinite simple groups" cannot be expressed in the language familiar to GAUSS. Even the word "group" would have meant to him, at best, not an abstract group, but a group of permutations. It certainly is more difficult to appropriate the meaning of a new concept than to learn a new word for a known one. All of this may very well have contributed to the popularity of mathematical conversations, seminars, expository lectures, etc. which we mentioned earlier. They offer, to many people, the easiest way to become familar with new concepts.

Chapter II.13

Geographical Distribution of Research and Effects of Migration

Our contribution to the closely related topics appearing in the heading of this chapter will be of a modest nature. Geographical distribution of mathematical research now comprises a much larger part of the globe than it did at the beginning of the Twentieth Century. India and Japan had already joined Europe and North America before World War I with important contributions to mathematics although not to our special field. Since then, contributions have come in increasing numbers from many more countries, and in a remarkable number of cases, from mathematicians who were natives of other countries. For instance, after World War I a rapidly increasing number of authors were natives of China, although they did not work there, at least not at the time of the publication of their papers. As it happens, our bibliography lists only one Chinese author, HUA, who at the time of the appearance of his paper mentioned there was in the United States but later returned to his native country.

We cannot discuss here the distribution of mathematical research in general or its spreading over larger areas of the globe, since we have to restrict our account to combinatorial group theory. Even so, we shall confine ourselves to the enumeration of a few facts and, with respect to effects of migration, to giving some statistical data and examples. Certainly, we shall not try to distill general principles out of the limited material we have to offer. Merely as an aside, we mention that none of the facts we shall submit contradict a general rule formulated by LIPMAN BERS in a talk given at Columbia University in the Fall of 1980, which states that research in a country will profit from the immigration of scholars only if it is already on a level comparable to that of the immigrants.

Until 1933, that is, for more than half of the time interval of the existence of combinatorial group theory, the overwhelming majority of papers in the field was written in German. We have no explanation for this fact. The important stimulant introduced by POINCARÉ through his discovery of the fundamental group of a space was taken up by TIETZE and DEHN but, in spite of the publication of the monograph by DE SÉGUIER, had no noticeable influence in France. The papers by HUMBERT, VOGT, and by the Italian, BIANCHI deal with marginal aspects of the theory. England, the country of CAYLEY and BURNSIDE, contributed to other aspects of group theory but not to the theory of group presentations. Particularly perplexing is the absence of American contributions. After all, group theory in general had been studied in America for a long time and with great success. Several textbooks on finite groups had been published, and hundreds of papers on this subject had been produced by the prolific G. A. MILLER and many other authors. The brilliant work on simple groups of finite order by L. E. DICKSON was, in 1933, an internationally recognized classic. Even the graph of a group had found attention. We have cited MASCHKE [1896], and this paper was followed later by other publications in American journals. But the theory of group presentations did not appear in American journals or in papers by American authors except for presentations of special groups which were used merely as a concise way of defining them.

Suddenly, in 1933–1934, there appeared papers by four completely unrelated authors who were instrumental in providing an international forum for combinatorial group theory. One of them is H. S. M. COXETER, whose three papers published in 1934 (listed in COXETER and MOSER [1973]) appeared in British and American Journals and systematically investigate groups generated by reflections and related topics. We discussed his work briefly in Chapter II.10. In 1933, VAN KAMPEN published his important papers (mentioned at the beginning of Chapter II.9) in the *American Journal of Mathematics*. The long range effect of the paper by PHILIP HALL, published in 1933 in England, was described in Chapter II.7. And in Chapter II.4, we have tried to indicate why the papers by KUROSH [1933 and 1934] mark the beginning of the phenomenal flourishing of combinatorial group theory in the Soviet Union. There is hardly any aspect of the field to which authors from that country have not made profound contributions. By way of explanation (or, perhaps, we should say, apology), we have to mention here that our bibliography does not reflect this fact adequately. The reason for this is a practical one. Our text has been written for English-speaking mathematicians, and whenever we had to refer to surveys rather than to individual papers, we chose, of course, those written in English. Together with Russian, these are now the dominant languages in the field. This does not mean that today (1980) the authors in the field come only from the Soviet Union or from English-speaking countries, although Great Britain, Canada, the United States, and Australia are strongly represented. Mathematicians from other countries frequently use English for their publications, although certainly mathematical papers in many other languages continue to appear. One may say that today research in combinatorial group theory is distributed rather widely, although not evenly, throughout the mathematical world. And its basic concepts, in particular, those of a free group and of a presentation, now appear in most textbooks on group theory. This was not at all the case in 1933. Going back to that year, we shall sketch some of the developments which led from that time to the present situation.

In the introductions to Chapters II.2–II.6, we mentioned the names of NIELSEN, REIDEMEISTER, SCHREIER, DEHN, ARTIN, and FURTWÄNGLER as those of leading contributors to the development of combinatorial group theory. By 1933, SCHREIER had been dead for four years. With the exception of REIDEMEISTER, the other mathematicians had lost interest in the field, at least temporarily. This meant that for several years to come, REIDEMEISTER was the only one of the mathematicians with an active interest in combinatorial group theory who had been trained in a German-speaking university and who also had a well-established reputation and a tenured academic position. Most of the other authors in this class were in their twenties and none of them was in a secure position. In fact, for several of them, the very opposite was true. They lost the positions they had held because 1933 was also the year in which HITLER came into power in Germany. Scholars who themselves or whose spouses were of Jewish descent were barred from academic positions. REINHOLD BAER went to England and from there to America. FRIEDRICH LEVI went to India. BERNHARD NEUMANN and KURT HIRSCH went to England. MAX DEHN was forced into early retirement and left Germany for NORWAY in 1939. ARTIN went to America in 1935. (As an aside, we mention that REIDEMEISTER was temporarily suspended from office and then transferred to a smaller university for being "politically unreliable.") As we shall see later, some of these migrations eventually helped the spread of interest in combinatorial group theory to other countries. Except for one case (Australia), we cannot say exactly how much. After all, the reading of literature also provided a channel for the transmission of ideas and problems at all times. KUROSH never met SCHREIER, but he read his paper. The same is true for MARSHALL HALL JR., whose 1938 paper is based on another part of SCHREIER's work. Also, combinatorial group theory was an offshoot of topology and continued to be a useful tool in this discipline. The interest in topology had already spread rapidly to many countries at an earlier time, and important contributions to combinatorial group theory continued to come from scholars who were primarily topologists. We mention here only a few: H. HOPF, J. H. C. WHITEHEAD, and S. EILENBERG.

Nevertheless, migration of scholars undoubtedly had its effects. These effects were blurred to an increasing degree through the introduction of visiting professorships and temporary membership in research institutes which started around 1930 and became very common after 1950, particularly through the extensive hospitality of American institutions. We shall exclude these arrangements from our considerations, restricting our attention to cases where a mathematician's home was in a country different from the one whose citizenship he or she held at the time of birth. In those cases where the training of the scholar was provided by the host country, the only (but by no means negligible) effect was an enrichment of the talent pool of that country. We are not quite sure how many of the names appearing in our bibliography should be listed in this class. Probably there are not more than six. As a clear-cut example, we mention E. L. POST (1897–1954) who was born in Augustow (at that time, Russia) and came to America at the age of seven. In addition, we have 32 scholars who changed their country of residence after they had begun to publish mathematical research of their own. It is those who, in many cases, contributed to the spread of interest in combinatorial group theory to other countries or were influential in other respects, enhancing the attractiveness of the institutions which they joined through their prestige or their activities as teachers or organizers. As an example of a scholar who exerted this type of influence whose name appears in our bibliography, although he did not contribute to combinatorial group theory, we mention HERMANN WEYL, who left Goettingen in 1933 and joined the Institute of Advanced Study in Princeton.

The total of 39 names of migrating scholars in our bibliography comprises a little less than one-sixth of all names. Of these, 21 were refugees, and 18 of these were refugees from HITLER (although not necessarily from Germany). The others came for a variety of reasons which include the cultural and scientific attractiveness of the host country as well as economic and even purely personal reasons. English-speaking countries (Australia, Canada, the United Kingdom, and the United States) played host in three-quarters of all cases. The remaining host countries were France, Germany, India, Israel, Norway, Switzerland, and the Soviet Union.

Even in the rather stable world before World War I, we have a few examples of migration. LIE went to Germany for several years. His stay there helped to spread interest in his work, but there can be no doubt that this effect was not a very strong one. LIE's ideas and results attracted attention independently of his stay in Leipzig in many parts of the mathematical world. MASCHKE joined the University of Chicago where he died a few years later before he could exert any recorded influence. His stay there had been arranged by FELIX KLEIN, who tried to persuade German mathematicians to emigrate to the United States, since he believed that this country would become increasingly important for mathematical research. That HURWITZ and, temporarily, MINKOWSKI went from Germany to Switzerland would, at that time, hardly have been considered a case of migration. However, we have, even before World War I, an example of a refugee. SCHUR, being Jewish, had no chance whatsoever of sustaining himself as a mathematician in czarist Russia and went to Germany. He shared with DEHN the fate of becoming a refugee a second time. A leading mathematician and an outstanding and highly successful teacher, he had occupied for 16 years the very prestigious chair at the University of Berlin. He escaped, with great difficulty, from HITLER's Germany in 1939 to Israel where he died without resuming an active life as a scholar. DEHN found asylum in Norway in 1939, but had to flee under dangerous circumstances in 1940 to America. Although he had some students there, he did not find the recognition that he deserved. His name is today more widely known in the English-speaking world than it was nearly 30 years ago at the time of his death.

So far, we have mentioned mainly the migration of scholars whose reputation had been established before World War I. With the exception of DEHN, none of them has played a more than marginal role in the history of combinatorial group theory and we have not recorded a single case where the effects of migration for the transmission of ideas could be shown to have been decisive, although other effects (contribution to the talent pool and to the attractiveness of academic institutions) have been demonstrated, e.g., by the names of SCHUR and WEYL. We turn now to case histories involving a later generation of mathematicians, starting with those of K. A. HIRSCH, B. H. NEUMANN, and HANNA NEUMANN (née VON KAEMMERER). They went (as refugees) from Germany to England in or shortly after 1933 and acquired second doctorates there, although B. H. NEUMANN (a student of I. SCHUR) and HIRSCH already held a Ph.D. from German Universities. HANNA NEUMANN had the equivalent of a Master's Degree. She, too, was strongly influenced by SCHUR.

We have mentioned the work of HIRSCH in Chapter II.10. Apart from his publications, HIRSCH influenced the development of mathematical research in England through his making the mathematics department at Queen Mary College at the University of London a center of research in algebra. As far as we can know, a department with a strong coherence of the research interests of its members was a novelty, at least in England. Until 1948, combinatorial group theory was represented in England mainly by HIRSCH, B. H. NEUMANN, WHITEHEAD, and, of course, P. HALL. In 1948, HANNA NEUMANN joined the development of the field which from then on grew rapidly. This was due in part to the influence of P. HALL which goes far beyond that of his own papers and was based also on the stimulating effect of his lectures. A large number of a whole generation of algebraists trained in England have been inspired by him. Of these, we mention KARL GRUENBERG (who, incidentally, also was a refugee, in this case, from Austria) and G. HIGMAN, whose paper written jointly with B. H. NEUMANN and HANNA NEUMANN is perhaps the most widely used publication in combinatorial group theory. However, the share of the work of HIRSCH and of the NEUMANNS in the development of combinatorial group theory is clearly visible. It is remarkable that our story does not end here. B. H. NEUMANN was also influential as a teacher, and one of his Ph.D. students was G. BAUMSLAG (a native of South Africa), who shortly after obtaining his doctorate, emigrated to the United States. His great share in the flourishing research in combinatorial group theory on the North American continent can be mentioned here only in passing. It is merely a part of the effect of the migration of the NEUMANNS, which continued with their acceptance, in 1960, of an invitation to join the Australian National University in Canberra in 1962–1963. In the coming years, Canberra developed into a center for group-theoretical research, attracting visitors, producing young scholars active in the field, and playing host to two international conferences, the proceedings of which were published as books under the following titles in 1967 and in 1974: *Proceedings of the International Conference on the Theory of Groups, Canberra, 1965* edited by B. H. Neumann and L. G. Kovács, Gordon and Breach, New York; *International Conference on the Theory of Groups* edited by M. F. Newman, Lecture Notes in Mathematics 372, Springer-Verlag, Berlin-Heidelberg-New York. The list of participants clearly shows the increase in the number of Australian scholars.

Of course, this success resulted not only from the presence of B. H. NEUMANN and HANNA NEUMANN in Canberra. It was made possible by a fortunate combination of circumstances: The right people in the right place were being supported by the financial and organizational contributions of the Australian National University.

The case of REINHOLD BAER (1902–1979) offers an example of three migrations. After 2 years at the University of Manchester, he went to America and became a faculty member of the University of Illinois in Urbana. In the years from 1940 to 1957, he had 20 Ph.D. students, many of whom continued to be active in research later on. Returning to Germany (the University at Frankfurt am Main) in 1957, he had within merely 11 years, 28 Ph.D. students. By all standards, this is a remarkable achievement. What makes it nearly (and, perhaps, absolutely) unique within mathematics is the fact that at least 25 of these continued to be active in research. Incidentally, the work of his Ph.D. students is not confined to group theory or even to algebra. Several of the thesis topics belong to geometry or analysis.

There are less conspicuous incidents of the transfer of research problems through migration. KALOUJNINE, a Soviet citizen, left Hamburg in 1933 and went to Paris where, after World War II, he met LAZARD as a student and suggested to him the study of the applications of Lie rings to combinatorial group theory, something KALOUJNINE was well acquainted with because he continued to keep abreast of the German literature. The result was the fundamental paper by LAZARD [1953].

We could continue providing more case histories, describing various degrees of scientific influence due to migration, going from none to considerable, but we believe that this would not make much sense for two reasons. Combinatorial group theory is too limited a field to provide a complete picture. And apart from this fact, there may be important influences which are due to migration and are of a more intangible nature than the transfer of interest and research activity in a special field from one country to another but which cannot easily be documented. The style of teaching and writing are important in every field; in mathematics, the emphasis on abstraction and axiomatics, or on motivation and historical context, or the establishing of connections between different fields are examples of phenomena which are subject to the influence of the migration of scholars. The same is true for the standards of training and education, including the planning of a curriculum. A single discipline of research is too narrow a basis for a study of all of these things. In addition, an examination of intangibles readily invites an amount of speculation from which we have tried to abstain in our report.

Chapter II.14

Organization of Knowledge

There exist two phenomena in the development of mathematics in the Twentieth Century which have no precedents in earlier times. They are the rates of increase in the volume of knowledge and in the complexity of proofs. We shall start with the first of these by reproducing below a statistical table which is due to W. R. UTZ of the University of Missouri and was published in the *Notices of the American Mathematical Society*, Issue 211, p. 424, August 1981. We have already mentioned this table in Chapter II.12 because of its two right-hand columns. What concerns us here are the two left-hand columns which show the number of titles of mathematical papers which appeared in the *Mathematical Reviews* during the last years of five consecutive decades.

Year	Number of titles	Number jointly authored	Percentage jointly authored
1940	1579	92	5.8
1950	3298	214	6.5
1960	4393	473	10.8
1970	12011	1680	14.0
1980	18383	3932	21.4

This table clearly exhibits the rapid growth of mathematical production in the decades after World War II. We have no exact data for the rate of growth of the literature on the theory of infinite groups (with the exclusion of Lie groups). Certainly, the production in this field is a very small part of the total production in mathematics. BAUMSLAG [1974] lists 4563 titles on infinite groups for the three decades 1940–1970. This number is only a little larger than that for all of mathematics in the single year 1960 which, in turn, is less than one-quarter of that for 1980.

The mere fact that the amount of documented mathematical knowledge exceeds the reading capacity of any individual human being is nothing new. It is doubtful that anybody during the past hundred years actually read all of the works of LEONHARD EULER or all of the 1742 papers on linear differential equations which appeared in the years 1865–1907 and are listed in the bibliography of the survey by SCHLESINGER [1909]. When MAGNUS [1939] completed his survey on "general group theory" (i.e., group theory with the exclusion of Lie groups and group representations), he estimated that the total literature on group theory at that time consisted of approximately 8000 papers. (About 4% of these were due to a single author, G. A. MILLER). Nobody ever read all of these either, and of course the overwhelming majority of all of these publications is now of interest mainly to historians of mathematics. Although no mathematical result can ever become truly obsolete—if it is true once, it continues to be true—mathematics has, at least since the Seventeenth Century, developed again and again powerful processes which make much of the work of earlier times dispensable. To use the examples mentioned at the beginning of this chapter, we observe that the papers by EULER on the quadratic law of reciprocity are numerous enough to make it difficult for a modern reader to ascertain the fact that EULER did *not* prove it in all cases. But not only did GAUSS prove it in its full generality; there now exists a very brief and lucid proof for it. Much of the results and the proofs of the theorems in SCHLESINGER's bibliography can now be derived from powerful existence and uniqueness theorems, and as far as the many special cases are concerned, it is enough to have a carefully classified survey for these which allows one to retrieve them quickly whenever they are needed. The same is true for much of the literature on group theory before, say, 1940. We shall characterize these two processes as *concentration* and *storage*. In combinatorial group theory, the Reidemeister–Schreier method or the Kurosh subgroup theorem for free products are examples of concentration, and the survey by COXETER and MOSER [1973] provides an example of storage. Neither one of the two processes can be said to be fully automatic. One has to learn how to apply a powerful and general theorem to special cases, and storage has to be based on an ordering principle which, in a rather undefined way, appears to be "natural" to the mind of a mathematician. Nothing like the alphabetical order of the words in a dictionary exists which would be truly helpful in retrieving mathematical knowledge from storage.

Before reporting on the ways in which concentration and storage have been developed further in order to cope with the enormous increase in mathematical knowledge during the past 50 years, we have to mention another problem which developed simultaneously: the complexity of proofs.

This complexity has two aspects. One of them is the vertical structure of mathematical disciplines which requires an understanding and appropriation of earlier results in order to understand or even to formulate the later ones. This aspect becomes increasingly important the older a mathematical discipline gets. The other aspect is based on the fact that even the proof of a single theorem may require very long chains or, even worse, "trees" of arguments, i.e. the consideration of many cases and subcases. Combinatorial group theory, being a rather young discipline, has, for a remarkably long time, not been affected very much by the first aspect, but rather early and very strongly by the second one. We shall now illustrate this observation with some examples.

The papers by TARTAKOWSKII [1949] as well as the even later papers by GREENDLINGER [1960a and 1960b] deal with what is now called "small cancellation theory" and provide a solution of the word problem for large classes of finitely presented groups. Although the word problem had been formulated already in 1910 by DEHN, and although it had been, at least since 1920, the topic of a fair number of investigations, practically no knowledge of any earlier paper is needed for the understanding of the work of TARTAKOWSKII or GREENDLINGER. On the other hand, their papers, in particular, the one by TARTAKOWSKII, offer examples of proofs with many cases and subcases and are of a very complex nature. They are by no means the first ones of this type in combinatorial group theory. As early as 1924, the paper by NIELSEN [1924a] is a 41-page paper with very complex and difficult proofs. Its content was described in Chapter II.2, together with the work of McCOOL [1974, 1975a and 1975b] which simplified and generalized NIELSEN's work. Not mentioned in Chapter II.2 is a paper by DE SÉGUIER [1924], which appeared shortly after NIELSEN's paper, consists of only four pages and proposes to solve the same problem; viz. to find a presentation for the automorphism groups of free groups of finite rank. It exemplifies an inadequate, or perhaps we should say, inadmissible method of coping with the complexity of proofs. DE SÉGUIER's paper is so condensed and sketchy that there arise serious doubts whether it is correct. MAGNUS tried to understand it in 1934 and, after giving up, asked NIELSEN his opinion. NIELSEN promptly replied that he had not been able to understand it either.

The situation where the correctness of a result is vouchsafed for a long time only by the reputation of a single scholar seems to be a novel one which did not arise before, or at least not long before, the Twentieth Century. (Here we are disregarding the much older phenomenon that proofs have to be abandoned as obsolete because the borderline between intuitive and rigorous arguments had to be redrawn.)

The very serious difficulties which are posed by the increasing complexity of proofs can be resolved only through a process which we shall call *streamlining*. It includes the simplification of proofs through clever artifices, new methods, and through the embedding of a particular theorem in the context of an encompassing theory in which it appears as one of many results and in which the steps from one result to the next are easily understood without overtaxing the attention span of the reader.

For much of small cancellation theory, the streamlining of proofs was begun shortly after the appearance of GREENDLINGER's paper in 1960 and has been documented in LYNDON and SCHUPP [1976]. That the streamlining of NIELSEN's [1924a] paper by McCOOL took place only after half a century may be explained in part by the fact that its result, although quoted in several surveys, has actually been used rather sparingly. We know of only three papers (B. H. NEUMANN [1932], MAGNUS [1934a], and Fouxe-Rabinovitch [1941]) in which it plays an essential role. On the other hand, the interest in the automorphism groups of free groups has always been maintained by their relation to the theory of mapping class groups of two-dimensional manifolds which was sketched very briefly in Chapter II.10 and which had attracted an increasing amount of attention in the decade before the publication of McCOOL's papers.

We shall now try to describe the efforts which have been made since World War I to overcome the difficulties arising from the proliferation of research and from the increasing complexity of papers, giving examples mainly from combinatorial group theory.

When, in 1927, the German encyclopedia of mathematical sciences had finally (after nearly 30 years) been completed, it was glaringly obvious that much of it was already obsolete. Preparations for a new edition started a few years later.

A few articles appeared before and after World War II which started in the same month in which the article on "general group theory" appeared. (This one, at least, was completed and did not break off in the middle like the corresponding article of the French encyclopedia which was interrupted by World War I.) Apart from the war, work on the second edition was also handicapped by the fact that the politics of HITLER's Germany made a truly international cooperation of mathematicians impossible. Even the first edition had used some non-German authors. For the second edition, this would have been a necessity even under the best of circumstances. With the greatly reduced capacity for research in Germany, due to the exodus of many of its scholars, the situation became hopeless. The result was a handful of articles which, temporarily at least, were of some use but were restricted in their effectiveness by another phenomenon. The old style encyclopedia itself showed signs of becoming an inadequate method for the organization of mathematics. It was supposed to provide a source and, if at all possible, a complete source of information about the development and factual knowledge of the individual disciplines. But the traditional style of encyclopedias excluded the description of methods and proofs and forced, or at least induced, the authors (with a few exceptions) to produce contributions which served only the purpose which we characterize as storage. This is helpful but not enough. Apart from this shortcoming, the main reason which seems to have made the idea of the old style encyclopedia obsolete was its rigidity. It required a plan and a parceling out of the various contributions. But mathematics began to change so rapidly that the plan was out of date long before it could be fully implemented. More flexible and more open-ended approaches were needed, and they began to emerge in the years between the two world wars. We shall now try to enumerate them.

Increasingly, the old-style textbook was being replaced by the monograph. A textbook was (and still is, in the case of elementary topics like calculus) a self-contained presentation of a topic. It may mention names but will cite few if any references. The proofs have been selected and worked out by the author. A monograph may have the same features, but it will also refer to the literature, mentioning methods and results not fully presented in the book and thereby facilitating access to the literature for the reader. The works of WEBER [1894–1896] and even the one by FRICKE [1926] are typical textbooks. On the other hand, BURNSIDE [1911] and KUROSH [1944] are monographs in our terminology, in spite of the fact that they are still largely self-contained. Also, a textbook addresses itself mainly to the student of mathematics, and a monograph is meant to be an aid to the research mathematician. However, there is no sharp line separating the two categories. Monographs may be used as textbooks for a course taught at a university, and, at least at the time when the work by WEBER [1894–1896] appeared, it was also of great interest to the active scholar. In any case, a large part of today's mathematical knowledge is being deposited in monographs, and they serve, above all, the purposes which we have called concentration of knowledge and streamlining of proofs.

Related to monographs are surveys which may take the form of a classified guide to the literature, serving mainly the purpose of storage, or may contain proofs or sketches of proofs and methods, serving also the purposes of concentration and streamlining. An example for the first type of survey is offered by COXETER and MOSER [1973]. The second type is well represented by LYNDON and SCHUPP [1976].

Certainly, neither monographs nor surveys are a novelty of the Twentieth Century, but they are not much older. And their importance has increased rapidly since about 1930 when they began to take over the functions of the encyclopedia. At that time, open-ended series of monographs and surveys began to appear with the purpose of keeping up with the development of mathematics. They were offered by commercial publishers, research institutes, mathematical societies, and university presses in many countries and supplemented by survey articles in journals and conference reports. Since 1978, the *Bulletin of the American Mathematical Society* publishes several such articles in every issue. A novel form of survey was introduced after 1970 by the editors of the *Mathematical Reviews*. It consists of collections of the reviews of mathematical papers which appeared during the years 1940–1970 for several fields. The reviews are arranged according to a classification of the topics which has been worked out by the editor and are supplemented by cross references to other reviews where the same paper or book has been mentioned. An index of the titles, ordered alphabetically according to the names of the authors, appears at the end of the work. This index is also available for papers reviewed in the *Zentralblatt der Mathematik* for the years 1930–1940. Merely reading the table of contents of such a survey gives a good idea of the structure of the field which it covers. For the authors of the present book, it is fortunate that combinatorial group theory is covered by the survey due to BAUMSLAG [1974]. We have used it extensively.

It cannot be denied that nearly all of the series of monographs and surveys in existence are of a more haphazard character than an encyclopedia. However, there exists one series of publications whose purpose is a systematic organization of mathematical knowledge. The name of the author on the title page is always the same: N. BOURBAKI. The way in which these publications are written has been described by HENRI CARTAN [1980] on the occasion of the 50th anniversary of this enterprise. We cannot describe its basic philosophy here and do not have a strong reason to do so, since combinatorial group theory has, at least so far, hardly been touched by BOURBAKI. The only one of its publications which we would consider as important for this field is BOURBAKI [1968]. For this reason, we also cannot assess the influence of BOURBAKI, except by mentioning that these publications have contributed to a standardization of terminology and even of notation in mathematics which, at least in part, seems to have been universally accepted. That this is a remarkable achievement becomes obvious if one notes that the concentration of mathematical knowledge has, in part, been paid for by an enormous proliferation of technical terms.

As an aside, we mention here a nonmathematical and, as we believe, very unusual aspect of the enterprise covered by the name N. BOURBAKI. Mathematicians, as a class, are known to be rather individualistic. Nevertheless, it is of course not unlikely that some of them should join in a common venture. The group theorists who worked on the enumeration of the simple groups of finite order are a current example. But here we have a common goal of a well-defined, limited, and purely mathematical nature. Working as a Bourbakist means subscribing to a definite (although, perhaps, not abstractly formulated) philosophy of mathematics. The task, like that of the Académie Française of protecting the French language, is open ended, but in contradistinction to the membership in the French Academy, working as a Bourbakist imposes great demands on the scholars' time, is done anonymously, and not rewarded by honors and prestige. Still, there are other instances of serendipity where like-minded people have joined in promoting a common cause of an intellectual nature. According to the 11th edition of the *Encyclopedia Britannica*, the Académie Française was founded in this manner. However, a few years later, it was stabilized by RICHELIEU, who made it a state institution. The unusual feature of Bourbaki is that it continues to function on the same precarious basis on which it was founded after half a century of profound changes in the structure and composition of the world of mathematics.

There arises the question how effective all of these efforts to organize mathematical knowledge have been. In order to gives a more precise meaning to this question, we shall formulate a standard for mathematical production which, at least in principle although certainly not always in practice, had tacitly been accepted by the mathematical community of the late Nineteenth and early Twentieth Century. According to this standard, one would not use a theorem proved by someone else without having understood its proof. We cannot analyze here the elusive question of what it means to understand a proof. Arguments in topology which would have been accepted as valid in 1900 might have been rejected as intuitive but not rigorous in 1950. But in broad areas of mathematics, this question has been settled since the end of the Nineteenth Century. And we have the example of the prominent mathematician EDMUND LANDAU (1877–1938) whose work, even in its most minute details, conforms with the standard set above. Of course, we are unable to produce evidence showing that LANDAU's standard cannot be maintained any more in, say, combinatorial group theory, but we should like to venture the opinion that this would today be more difficult in this field than it was in number theory at the time of LANDAU. In support of our opinion we mention the omission of calculations (which then are called "straightforward" or "routine") in many research papers and monographs. Here we even disregard the rather novel and delicate question of the validity of evidence which has been provided only by using an electronic computer.

Whereas a relaxation of the standard represented by LANDAU might be a reason for serious concern, the question of duplication of results in the literature, although much less disturbing, could also be used as a test for the effectiveness of the modern methods of organization of knowledge.

Clearly, even the most careful reviewing procedures and the most lucidly organized monographs and surveys cannot entirely exclude the duplication of results, even if we confine this term to cases where the same method has been used to obtain the same theorem. Browsing through the *Mathematical Reviews* where duplications are mentioned if they are observed by the reviewer, we receive the impression that this extreme case of duplication is not particularly common, at least not in relation to the total volume of new results. Even the cases where a paper produces a result that turns out to be a special case of an earlier theorem are not very numerous. And these cases have at least the merit of enhancing the value of the more general result.

Bibliography

Note: At the end of every entry, we list the chapters in which the reference has been quoted. This is done by using the Roman letters I and II, referring, respectively, to Part I and Part II, followed by the number of the chapters in those parts in Arabic digits. For instance, I.1, 2, 6.B, 8 means Part I, Chapters 1, 2, 8, and Section 6.B.

ADELSBERGER, H., 1930, Über unendliche diskrete Gruppen. *J. reine u. angewandte Math.* **163**, 104–124. I.6.E, II.7.

ADJAN (ADIAN, ADYAN), S. I., 1970. Infinite irreducible systems of group identities, *Izv. Akad. Nauk. SSSR. Ser. Mat.*, **34**, 715–734 (Russian). English translation in *Math. USSR — Izv.* **4**, 721–739, 1971. II.8.

ADJAN, S. I., 1975, *The Burnside Problem and Identities in Groups* (Russian), Izdat. Nauka, Moscow, 335 pp. English translation: *Ergebnisse der Mathematik und ihrer Grenzgebiete* 95, 1979, Springer-Verlag, Berlin-Heidelberg-New York. I.6.F, II.7.

ADO, I. D., 1936. Über die Darstellung von Lieschen Gruppen durch lineare Substitutionen, *Bull. Soc. Phys.-Math. Kazan* **111**, Ser. 7, 3–41 (Russian, German Summary). II.7.

ADO, I. D., 1947, The representation of Lie algebras by matrices. *Uspehi Mat. Nauk. (N.S.)* **2**, No. 6 (**22**) 159–173 (Russian). English translation in *American Math. Soc. Transl.* **2**, 1949. II.7.

ALEXANDER, J. W., 1928, Topological invariants of knots and links, *Trans. American Math. Soc.* **30**, 275–306. II.6.

ARTIN, E., 1925, Theorie der Zöpfe, *Abh. Math. Sem. Hamburg. Univ.* **4**, 47–72. II.3, 10.

ARTIN, E, 1926, Das freie Produkt von Gruppen, §92 in: F. Klein, *Vorlesungen über höhere Geometrie*, Julius Springer, Berlin, 1926. Reprint 1949: Chelsea, New York. I.2, II.4, 6.

ARTIN, E., 1927, Beweis des allgemeinen Reziprozitätsgesetzes, *Abh. Math. Sem. Hamburg Univ.* **5**, 353–363. II.6.

ARTIN, E., 1947b, Theory of braids, *Ann. of Math.* (*2*) **48**, 101–126. II.10.

ARTIN, E., 1947c, Braids and permutations, *Ann. of Math.* (*2*) **48**, 643–649. II.10.

ARTIN, E., 1965, *The Collected Papers of Emil Artin*, edited by S. Lang and J. T. Tate, 560 pp., Addison-Wesley, Reading, MA.

ARTZY, R., 1972, Kurt Reidemeister, *Jber. Deutsche Math.-Verein.* **74**, 96–104. II.3.

AUSLANDER, L., 1967, On a problem of Philip Hall, *Annals of Math.* (*2*) **86**, 112–116. II.6, 10.

AUSLANDER, L. and R. TOLIMIERI, 1975, Abelian harmonic analysis, theta functions and function algebras on a nilmanifold, 99 pp., *Lecture Notes in Mathematics* 436, Springer-Verlag, Berlin-Heidelberg-New York. II.10.

BACHMUTH, S., 1965 and 1966, Automorphisms of free metabelian groups, *Trans. American Math. Soc.* **118**, 93–104. Induced automorphisms of free groups and free metabelian groups, *Trans. American Math. Soc.* **122**, 1–17. II.6.

BAER, R., 1927, Kurventypen auf Flächen, *J. reine u. angew Math.* **156**, 231–246. II.4.

BAER, R., 1929, Die Abbildungstypen-Gruppe der orientierbaren geschlossenen Flächen vom Geschlecht 2, *J. reine u. angew. Math.* **160**, 1–25. II.4.

BAER, R., 1934, Erweiterung von Gruppen und ihren Automorphismen, *Math. Z.* **38**, 375–416. II.9.

BAER, R. and F. LEVI, 1936, Freie Produkte und ihre Untergruppen, *Comp. Math.* **3**, 391–398. II.4.

BAER, R., 1945, Representations of groups as quotient groups I, II, III, *Trans. American Math. Soc.* **58**, 295–347, 348–389, 390–419. II.5, 10.

BAER, R., 1949, Groups with descending chain conditions for normal subgroups, *Duke Math. J.* **16**, 1–22. II.10.

BAER, R., 1952, *Linear Algebra and Projective Geometry*, 318 pp., Academic Press, New York. II.4.

BAGHERZADEH, G. H., 1975, Commutativity in one-relator groups, *J. London Math. Soc.* **13**, 459–471. II.5.

BAKER, H. F., 1904, Alternants and continuous groups, *Proc. London Math. Soc.* (2) **3**, 24–47. I.6.E, II.7.

BASS, H., MILNOR, J., and SERRE, J. P., 1967, *Solution of the congruence subgroup theorem for* SL_n *(n ≥ 3) and* Sp_{2n} *(n ≥ 2)*, Inst. Hautes Etudes Sci. Publ. Math. No. 33. I.6B, II.2.

BAUMSLAG, G., 1959, Wreath products and p-groups, *Proc. Cambridge Philos. Soc.* **55**, 224–231. II.8.

BAUMSLAG, G., 1960, Some aspects of groups with unique roots, *Acta Math.* **104**, 217–303. II.5.

BAUMSLAG, G. and D. SOLITAR, 1962, Some two-generator, one-relator nonhopfian groups, *Bull. American Math. Soc.* **68**, 199–201. II.5.

BAUMSLAG, G., 1963, Automorphism groups of residually finite groups, *J. London Math. Soc.* **38**, 117–118. II.2.

BAUMSLAG, G., 1969, Groups with the same lower central sequence as a relatively free group II. Properties, *Trans. American Math. Soc.* **142**, 507–538. II.7.

BAUMSLAG, G., 1971, A finitely generated infinitely related group with trivial multiplicator, *Bull. Australian Math. Soc.* **5**, 131–136. II.9.

BAUMSLAG, G., 1973, Finitely presented metabelian groups. *Proc. 2nd Internat. Conf. Theory of Groups, Canberra*, 1973, 65–74. Lecture Notes in Mathematics 372, Springer-Verlag, Berlin-Heidelberg-New York. II.6.

BAUMSLAG, G., 1974, *Reviews on Infinite Groups*, 2 Volumes, American Math. Soc., Providence, RI. II.5, 6, 7, 8, 9, 10, 11.

BAUMSLAG, G., 1976a, Multiplicators and metabelian groups, *J. Australian Math. Soc.* (*Series A*) **22**, 305–312. I.6.F, II.9.

BAUMSLAG, G., 1976b, A remark on groups with trivial multiplicator, *American J. Math.* **97**, 863–864. I.6F, II.9.

BEHR, H., 1975, Eine endliche Präsentation der Gruppe $Sp_4(\mathbb{Z})$, *Math. Z.* **141**, 47–56. I.6A.

BERS, L. and I. KRA, Editors, 1974, A crash course on Kleinean groups, 129 pp., *Lecture Notes in Mathematics* 400, Springer-Verlag, Berlin-Heidelberg-New York. II.10.

BETTI, E., 1871, Sopra gli spazi di un numero qualunque di dimensioni, *Annali Mat. Puri ed applicati* (2) **4**, 140–158. II.9.

BIANCHI, L., 1892, Sui gruppi di sostituzioni lineari con coefficienti appartenenti a corpi quadratici imaginari, *Math. Annalen* **40**, 332–421. I.3, 6.A, B.

BIANCHI, L., 1893a, Sui gruppi di sostituzioni lineari, *Math. Annalen* **42**, 30–57. I.6.B.

BIANCHI, L., 1893b, Sopra alcune classi di gruppi di sostituzioni lineari a coefficienti complessi, *Math. Annalen* **43**, 101–135. I.6.B.

BIERBERBACH, L., 1911, Die Bewegungsgruppen der euklidischen Räume, *Math. Annalen* **70**, 297–336. I.6.A.

BIRKHOFF, G., 1935, On the structure of abstract algebras, *Proc. Cambridge Philos. Soc.* **31**, 433–454. II.8.

BIRKHOFF, G., 1937, Representability of Lie algebras and Lie groups by matrices, *Annals Math.* (2) **38**, 526–532. II.7.

BIRKHOFF, G., 1942, Lattice ordered groups, *Annals of Math.* (*2*), **43**, 298–331. II.6.

BIRKHOFF, G., 1946, Review of Everatt and Ulam: "On ordered groups", *Math. Reviews* **7**, 4. II.6.

BIRMAN, J. S., 1973, Poincaré's conjecture and the homeotopy group of a closed, orientable 2-manifold. *J. Australian Math. Soc.* **17**, 214–221. I.4.

BIRMAN, J. S., 1975, *Braids, Links and Mapping Class Groups*, Princeton University Press, Princeton, NJ (Annals of Math. Studies, No. 82). I.6A, D. II.10.

BLUMENTHAL, O., 1903 and 1904, Über Modulfunctionen von mehreren Veränderlichen, *Math. Annalen* **56**, 509–547; **58**, 497–527.

BOHNENBLUST, F., 1947, The algebraic braid group, *Annals of Math.* (*2*), **48**, 127–136. II.10.

BOONE. W. W., 1955, Certain simple, unsolvable problems of group theory IV, V, VI, *Nederl. Akad-Wetensch. Proc. Ser. A.* **58** (Indag. Math. **17**, 571–577; **19**, 22–27, 227–232). II.11.

BOONE, W. W. and G. HIGMAN, 1974, An algebraic characterization of the unsolvability of the word problem, *J. Australian Math. Soc.* **18**, 41–53. II.11.

BOURBAKI, N., 1968, *Groupes et Algèbres de Lie*, Chaps. 4, 5, 6, 288 pp., Hermann, Paris. II.10, 14.

BOY, W., 1903, Über die Curvatura integra und die Topologie geschlossener Flächen, *Math. Annalen* **57**, 151–184. (Report on dissertation with the same title.) I.8.

BRAUNER, K., 1928, Zur Geometrie der Funktionen zweier komplexer Veränderlicher, *Abh. Math. Seminar Hamburg. Univ.* **6**, 1–55. I.4.

BRITTON, J. L., 1963, The word problem, *Annals of Math.* **77**, 16–32. II.4.

BURAU, W., 1932, Über Zopfinvarianten, *Abh. Math. Sem. Hamburg Univ.* **9**, 117–124. II.10.

BURAU, W., 1934, Kennzeichnung der Schlauchverkettungen, *Abh. Math. Sem. Hamburg. Univ.* **10**, 285–297. II.10.

BURAU, W., 1935, Über Verkettungsgruppen, *Abh. Math. Sem. Hansischen Univ.* **11**, 171–178. II.10.

BURNSIDE, W., 1897a, *Theory of Groups of Finite Order*, Cambridge University Press, Cambridge, England. I.2.

BURNSIDE, W., 1897b, Note on the symmetric group, *Proc. London Math. Soc.* (*1*) **28**, 119–129. I.2.

BURNSIDE, W., 1899, Note on the simple group of order 504, *Math. Annalen* **52**, 174–176. I.2.

BURNSIDE, W., 1902, On an unsettled question in the theory of discontinuous groups, *Quart. J. Math.* **33**, 230–238. I.6.F.

BURNSIDE, W., 1911, *Theory of Groups of Finite Order*, 2nd ed., 512 pp. Cambridge University Press, Cambridge, England. Reprint: Dover, New York, 1955. I.2, 5, II.7.

BURNSIDE, W., 1912 and 1913, On some properties of groups whose orders are powers of a prime, I, II, *Proc. London Math. Soc.* **11**, 225–245; **13**, 6–12. II.7.

BURROW, M. D., 1958, Invariants of free Lie rings, *Comm. Pure and Appl. Math.* **11**, 419–431. II.7.

CAMM, RUTH, 1953, Simple free products, *J. London Math. Soc.* **28**, 66–76. II.11.

CAMPBELL, J. E., 1898, On a law of combination of operators, *Proc. London Math. Soc.* **29**, 14–32. II.7.

CANTOR, G., 1898, Beiträge zur Begründung der transfiniten Mengenlehre, *Art. I. Math. Annalen* **46**, 481–512. I.2.

CARTAN, E., 1936, Sur les domaines bornées homogènes de l'espace de *m* variables complexes, *Abh. Math. Seminar Hansischen Univ.* **11**, 116–162. I.6.A.

CARTAN, HENRI, 1980, Nicolas Bourbaki and contemporary mathematics, *The Mathematical Intelligencer* **2**, 175–180. II.14.

CAYLEY, A., 1878a, On the theory of groups, *American J. Math.* **1**, 50–52. Reprinted in: *The Collected Papers of Arthur Cayley*, Volume 10, 401–403, Cambridge University Press, Cambridge, England, 1896. I.7.

CAYLEY, A., 1878b, On the theory of groups, *Proc. London Math. Soc.* **9**, 126–133. Reprinted in: *The Collected Papers of Arthur Cayley*, Volume 10, 324–330, Cambridge University Press, Cambridge, England, 1896. I.5.

CAYLEY, A., 1889, On the theory of groups, *American J. Math.* **11**, 139–157. Reprinted in: *The Collected Papers of Arthur Cayley*, Volume 12, 639–656, Cambridge University Press, Cambridge, England, 1897.

ČERNIKOV, S. N., 1940, Infinite locally solvable groups (Russian), *Mat. Sbornik* **7**, 35–64. II.6.

CHOW, KEI-LIANG, 1948, On the algebraic braid group, *Ann. of Math.* (2) **49**, 654–658. II.10.

CLEBSCH, A. and P. GORDAN, 1866, *Theorie der Abelschen Functionen* (pp. 305–306) Teubner, Leipzig. I.6A, 7.

COHN, P. M., 1968, A presentation for SL_2 for euclidean imaginary number fields, *Mathematica* **15**, 156–163. I.6.B.

COXETER, H. S. M. and W. O. MOSER, 1965 and 1972, *Generators and Relations for Discrete Groups*, 2nd ed., 161 pp., Ergebnisse der Mathematik 14. Springer-Verlag, New York-Berlin-Heidelberg. 3rd ed. 1972. I.2, 5, II.10, 14.

CROWELL, R. H. and R. H. FOX, 1963, *Introduction to Knot Theory*, 182 pp., Ginn, Boston. II.3, 10.

DANTZIG, D. VAN, 1932, Zur topologischen Algebra I. Komplettierungstheorie, *Math. Annalen* **107**, 587–626. II.9.

DEHN, M., 1901, Über den Rauminhalt. *Math. Annalen* **55**, 465–478. I.4.

DEHN, M. and P. HEEGAARD, 1907, *Analysis Situs*, Enzyklopädie der mathematischen Wissenschaften 3 (III AB3), pp. 153–220. Teubner, Leipzig-Berlin. I.4, 7, 8.

DEHN, M., 1910, Über die Topologie des dreidimensionalen Raumes, *Math. Annalen* **69**, 137–168. I.4, 5.

DEHN, M., 1911, Über unendliche diskontinuierliche Gruppen, *Math. Annalen* **71**, 116–144. I.4, 5. II.3.

DEHN M., 1912, Transformation der Kurven auf zweiseitigen Flächen, *Math. Annalen* **72**, 413–421. I.4, 7.

DEHN, M., 1914, Die beiden Kleeblattschlingen. *Math. Annalen* **75**, 1–12. I.4, 5.

DEHN, M., 1928, Über die geistige Eigenart des Mathematikers, Frankfurter Universitätsreden No. 27. 25 pp., Universitätsdruckerei Werner und Winter, Frankfurt am Main. Preface and I.5.

DEHN, M., 1938. Die Gruppe der Abbildungsklassen, *Acta Math.* **69**, 135–206. II.10.

DEHN, M. (See also SCHOENFLIES, A.)

DE SÉGUIER, J.-A., 1904, *Théorie des Groupes Finis. Eléments de la Théorie des Groupes Abstraits*, 176 pp., Gauthier Villars, Paris. I.2, 6.F.

DE SÉGUIER, J.-A., 1912, *Théorie des Groupes Finis. Eléments de la Théorie des Groupes de Substitutions*, 228 pp., Gauthier Villars, Paris. I.6.A, E.

DE SÉGUIER, J.-A., 1924, Sur les automorphisms de certaines groupes, *Comptes rendues hebdomadaires des séances de l'académie des sciences* **179**, 139–142. II.14.

DICKSON, L. E., 1901a, *Linear Groups with an Exposition of the Galois Field Theory*, 312 pp., Teubner, Leipzig. Reprinted: Dover, New York, 1958. I.6.A, E.

DICKSON, L. E., 1901b, Theory of linear groups in an arbitrary field, *Trans. American Math. Soc.* **2**, 383–391. I.6.E. II.7, 11.

DICKSON, L. E., 1905, A new system of simple groups, *Math. Annalen* **60**, 137–150. I.6.E. II.7.

DIETZMAN, A. P., A. KUROSCH, and A. T. UZKOW, 1938, Sylowsche Untergruppen von unendlichen Gruppen, *Recueil Mathématique* (*Math. Sbornik*) **3** (45) 179–185. II.4.

DRILLICK, A., 1971, The Picard Group, Ph.D. Thesis, New York University, New York. I.6.B.

DYCK, W. VON, 1882, Gruppentheoretische Studien (with 3 plates of drawings at end of volume), *Math. Annalen* **20**, 1–44. I.2, 6.B, 7. II.5, 7, 11.

DYCK, W. VON, 1883, Gruppentheoretische Studien II, *Math. Annalen* **22**, 70–108. I.2, 8.

DYER, J. L. and E. FORMANEK, 1975, The automorphism group of a free group is complete, *J. London Math. Soc.* (2) **11**, 181–190. II.2.

DYER, J. L. and P. SCOTT, 1975, Periodic automorphisms of free groups, *Comm. Algebra* **3**, 195–201. II.2.

EASTON, B. S., 1902, *The Constructive Development of Group Theory* (with bibliography), 88 pp., Boston. I.6.F.

EDWARDS, H. M., 1977, *Fermat's Last Theorem. A Genetic Introduction to Algebraic Number*

Theory, Springer-Verlag, New York-Heidelberg-Berlin. II.6.

EILENBERG, S. and S. MACLANE, 1949, Cohomology theory in abstract groups, *Annals of Math.* **48**, 51–78, 326–341. I.6.F. II.9.

Enzyklopädie der mathematischen Wissenschaften mit Einschluss ihrer Anwendungen, 1904–1927, Teubner, Leipzig, Berlin. I.8. II.14.

ERDÉLYI, A. (ed.), 1955, *Higher Transcendental Functions*, Vol. 3, McGraw-Hill, New York. II.7.

FEDERER, H. and B. JÓNSSON, 1950, Some properties of free groups, *Trans. American Math. Soc.* **68**, 1–27. II.2.

FEIT, W. and J. G. THOMPSON, 1963, Solvability of groups of odd order, *Pacific J. of Math.* **13**, 755–1029. I.6.F.

FENCHEL, W., 1960, Jakob Nielsen in Memoriam, *Acta Math.* **103**, Issues 3 and 4, pp. VII–XIX. II.2.

FINE, B., 1974, The structure of $\mathrm{PSL}_2(R)$; R, the ring of integers in a Euclidean quadratic imaginary number field. In: *Discontinuous Groups and Riemann Surfaces*, pp. 145–170, Annals of Mathematics Studies. No. 79, Princeton University Press, Princeton, NJ. I.6.B.

FITE, W. B., 1906, Groups whose orders are powers of a prime, *Trans. American Math. Soc.* **7**, 61–68. II.7.

FORMANEK, E. (See DYER, J. L.)

FOUXE-RABINOVITCH, D. I., 1940, Über die Nichteinfachheit einer lokal freien Gruppe., *Rec. Math. (Mat. Sbornik)* N.S. **7** (49) 327–328 (Russian, German Summary). II.10.

FOUXE-RABINOVITCH, D. I. (FUCHS-RABINOWITSCH, D. J.), 1941, Über die Automorphismengruppen der freien Produkte II, *Mat. Sbornik N.S.* **9** (51) 183–220 (Russian: German Summary). II.14.

FOX, R. H., 1953, Free differential calculus I, *Annals Math.* (2) **57**, 547–560. II.6.

FOX, R. H. *et al.* 1954 to 1960, Free differential calculus II, III, IV, V, *Annals Math* (2) **59**, 196–210; **64**, 407–419; **68**, 81–95; **71**, 408–422. II.6.

FRATTINI, G., 1885, Intorno alla generazioni dei gruppi di operazioni, *Atti Accad. naz. Lincei, Rend. Ivs.* **1**, 281–285. I.7.

FREUDENTHAL, H., 1931, Über Enden topologischer Räume und Gruppen., *Math. Z.* **33**, 692–713. II.9.

FREUDENTHAL, H., 1945, Über die Enden diskreter Räume und Gruppen, *Comment. Math. Helvetici* **17**, 1–38. II.9.

FRICKE, R., 1886, Über die Substitutionsgruppen welche zu den aus dem Legendreschen Integralmodul gezogenen Wurzeln gehören, *Math. Annalen* **28**, 99–118. I.6.B.

FRICKE, R. and F. KLEIN, 1897, Vorlesungen über die Theorie der automorphen Functionen, Vol. 1., 634 pp., Teubner, Leipzig, Berlin. Johnson Reprint, Academic Press, New York, 1966. I.3, 5, 6.A,B,D,E,F. II.10.

FRICKE, R., 1899, Über eine einfache Gruppe von 504 Operationen, *Math. Annalen* **52**, 321–339. I.2.

FRICKE, R. and F. KLEIN, 1912, Vorlesungen über die Theorie der automorphen Funktionen, Vol. 2, 668 pp., Teubner, Leipzig, Berlin. Johnson Reprint, Academic Press, New York, 1966. I.6.D, E.

FRICKE, R., 1913, Automorphe Funktionen mit Einschluss der elliptischen Modulfunktionen, *Enzyklopädie der Mathematischen Wissenschaften* II.B.4., pp. 349–470, Teubner, Leipzig, Berlin. I.6.B.

FRICKE, R., 1926, *Lehrbuch der Algebra.*, Vol. 2, Vieweg & Sohn, Braunschweig. I.6.B.

FRICKE, R. (See also KLEIN, F.)

FROBENIUS, F. G. and L. STICKELBERGER, 1879, Über Gruppen mit vertauschbaren Elementen. *J. reine u. angew. Math.* **86**, 217–262. Reprinted as No. 19 in Vol. 1 of *Ferdinand Georg Frobenius. Gesammelte Abhandlungen*, ed. J.-P. Serre, Springer-Verlag, Berlin-Heidelberg-New York, 1968. I.7. II.6, 7.

FROBENIUS, F. G., 1895, *Über endliche Gruppen*, Sitzungsberichte d. K. Pressischen Akademie d. Wissensch, Berlin, pp. 81–112, Jahrgang, 1895. Reprinted in: *Ferdinand Georg Frobenius, Gesammelte Abhandlungen*, Vol. 2., ed. J.-P. Serre, Springer-Verlag, Berlin-Heidelberg-New York, 1968. I.6.F.

FROBENIUS, F. G., 1898, *Über die Beziehungen zwischen den Charakteren einer Gruppe und denen ihrer Untergruppen*, Sitzungsberichte d. K. Preussischen Akademie d. Wissensch, Berlin, pp. 501–515, Jahrgang, 1898. Reprinted in: *Ferdinand Georg Frobenius. Gesammelte Abhandlungen*, Vol. 3, ed. J.-P. Serre, Springer-Verlag, Berlin-Heidelberg-New York, 1968. I.6.F.

FROBENIUS, F. G., 1898, *Über die Composition der Charaktere einer Gruppe*, Sitzungsberichte d. K. Preussischen Akademie d. Wissensch, Berlin, pp. 330–339, Jahrgang, 1899. Reprinted in: *Ferdinand Georg Frobenius. Gesammelte Abhandlungen*, Vol. 3, No. 58, ed. J.-P. Serre Springer-Verlag, Berlin-Heidelberg-New York, 1968. I.6.F.

FROBENIUS, F. G., 1911, *Über die unzerlegbaren diskreten Bewegungsgruppen*, Sitzungsberichte d. K. Preussischen Akademie d. Wissensch, Berlin, pp. 654–665, Jahrgang, 1911. Reprinted in: *Ferdinand George Frobenius. Gesammelte Abhandlungen*, Vol. 3, ed. J.-P. Serre, Springer-Verlag, Berlin-Heidelberg-New York, 1968. I.6.A.

FURTWÄNGLER, PH., 1930, Beweis der Hauptidealsatzes für den Klassenkörper algebraischer Zahlkörper, *Abh. Math. Sem. Hamburg. Univ.* **7**, 14–36. II.3, 6.

GASCHÜTZ, W., 1956, Erzeugendenzahl und Existenz von p-Faktorgruppen, *Archiv d. Math.* **7**, 91–93. I.6.F.

GERSTENHABER, M. and O. S. ROTHAUS, 1962, The solution of sets of equations in groups, *Proc. Nat. Acad. Sci. U.S.A.* **48**, 1531–1533. II.5, 11.

GIESEKING, H., 1912, Analytische Untersuchungen über topologische Gruppen, Ph.D. thesis, Münster i. W. (Germany), privately printed. (A copy exists at the University of Frankfurt am Main, Federal Republic of Germany.) I.5, 6.B.

GOLOD, E. S. and I. R. ŠAFARCVIČ, 1964, On the class field tower, *Isv. Akad. Nauk. SSSR. Ser. Mat.* **26**, 261–272 (Russian). English translation in *American Math. Soc. Transl. Ser. 2*, **48**, 91–106. (Contains some errors, e.g., reference [4].) II.6.

GOLOVIN, O. N., 1950, Nilpotent products of groups, *Mat. Sbornik N.S.* **27** (69), 427–454 (Russian). English translation in *American Math. Soc. Transl. Ser. 2*, **2**, 89–116 (1956). II.4.

GORDAN, P. (See CLEBSCH, A.)

GRAVÉ, D.-A., 1908, *Teorie Konetschnych Grupp* (Russian), Kiev.

GREENDLINGER, M., 1960a, Dehn's algorithm for the word problem, *Comm. Pure and Appl. Math.* **13**, 67–83. I.4, 9. II.14.

GREENDLINGER, M., 1960b, Dehn's algorithm for the conjugacy and word problems, with applications, *Comm. Pure and Appl. Math.* **13**, 641–677. I.4, 5, 9. II.14.

GROSSMAN, E. K., 1974, On the residual finiteness of certain mapping class groups, *J. London Math. Soc.* **9**, 160–164. II.2.

GRÜN, O., 1937, Über eine Faktorgruppe freier Gruppen I, *Deutsche Math.* **1**, 772–782. II.7.

GRÜN, O., 1940, Zusammenhang zwischen Potenzbildung und Kommutatorbildung, *J. f. d. reine u. angew. Math.* **182**, 158–177. II.7.

GRUENBERG, K. W., 1957, Residual properties of finite soluble groups, *Proc. London Math. Society (3)* **7**, 29–62. I.6.A. II.9.

GRUSCHKO (GRUSHKO), I., 1940, Über die Basen eines freien Produktes von Gruppen, *Rec. Math. (Mat. Sbornik)* N.S. **8** (50), 169–182 (Russian, German Summary). II.4.

GUPTA, K. G. and N. D. GUPTA, 1978, Generalized Magnus embeddings and some applications, *Math. Z.* **160**, 75–87. II.6.

HAKEN, W., 1962, Über das Homöomorphieproblem der 3-Mannigfaltigkeiten I, *Math. Z.* **80**, 89–120. II.10.

HAKEN, W., 1973, Connections between topological and group theoretical decision problems, pp. 427–441 in: *Word Problems*, ed. Boone–Cannonito–Lyndon, North Holland, Amsterdam. II.10.

HALANAY, A., 1947, La théorie de Galois des extensions séparables infinies et les groupes topologiques, *Bull. Math. Soc. Roumaine Sci.* **48**, 65–76. II.9.

HALL, M., 1938, Group rings and extensions, *Annals of Math.* **39**, 220–234. II.6, 9.

HALL, M. and T. RADÓ, 1948, On Schreier systems in free groups, *Trans. American Math. Soc.* **64**, 386–408. II.2.

HALL, M., 1949a, Coset representations in free groups, *Trans. American Math. Soc.* **67**, 421–432. II.2, 7.

HALL, M., 1949b, Subgroups of finite index in a free group, *Canadian Math. J.* **1**, 187–190. II.7.

HALL, M., 1950a, A topology for free groups and related groups, *Annals of Math.* (*2*) **52**, 127–139. II.5, 9.

HALL, M., 1950b, A basis for free Lie rings and higher commutators in free groups, *Proc. American Math. Soc.* **1**, 575–581. II.7.

HALL, M., 1958, Solution of the Burnside problem for exponent six, *Illinois J. of Math.* **2**, No. 4B (Miller Memorial issue), 764–786. II.7.

HALL, M., 1959, *The Theory of Groups* (Lemma 15.51, p. 231), 434 pp., Macmillan, New York. II.6, 9.

HALL, P., 1933, A contribution to the theory of groups of prime power order, *Proc. London Math. Soc.* (*3*) **36**, 29–95. I.6.E. II.7, 13.

HALL, P., 1936, The Eulerian functions of a group, *Quart. J. Math. Oxford Ser.* **7**, No. 26, 134–151. II.3, 7.

HALL, P., 1954a, Finiteness conditions for soluble groups, *Proc. London Math. Soc.* (*3*) **4**, 419–436. II.6.

HALL, P., 1954b, The splitting properties of relatively free groups, *Proc. London Math. Soc.* (*3*) **4**, 343–356. II.9.

HAMILTON, W. R., 1856a, Account of the isocian calculus, *Proc. Royal Irish Acad.* **6**, 415–16 (Vol. 6 appeared over the period 1853–1858. His paper was read on November 10, 1856.) Reprinted in Vol. 3 of: *The Mathematical Papers of Sir William Rowan Hamilton*, No. 54., p. 609, Cambridge University Press, Cambridge, England, 1967. I.8.

HAMILTON, W. R., 1856b, Memorandum respecting a new system of roots of unity, *Phil. Mag.* **12**, 496. Reprinted in: *The Mathematical Papers of Sir William Rowan Hamilton*, Vol. 3 (Algebra), No. 55, p. 610, Cambridge University Press, Cambridge, England, 1967. I.8.

HAUSDORFF, F., 1906, Die symbolische Exponentialformel in der Gruppentheorie, *Berichte d. Sächsischen Akad. d. Wissensch.* (*Math. Phys. Kl*) *Leipzig* **58**, 19–48. I.6.E. II.7.

HEEGAARD, P. (See DEHN, M.)

HIGGINS, P. J. and LYNDON, R. C., 1962 and 1974, Equivalence of elements under automorphisms of a free group, *J. London Math. Soc.* **8**, 254–258 (Preprints 1962 as Lecture Notes, Queen Mary College, London.) II.2.

HIGMAN, G., B. H. NEUMANN, and H. NEUMANN, 1949, Embedding theorems for groups, *J. London Math. Soc.* **24**, 247–254. II.4, 13.

HIGMAN, G., 1951a, A finitely generated infinite simple group, *J. London Math. Soc.* **26**, 61–64. II.3, 4, 11.

HIGMAN, G., 1951b, A finitely related group with an isomorphic proper quotient group, *J. London Math. Soc.* **26**, 59–61. II.5.

HIGMAN, G., 1961, Subgroups of finitely presented groups, *Proc. Royal Soc. London. Ser. A.*, 455–475. II.4, 11.

HIGMAN, G., 1974, *Finitely Presented Infinite Simple Groups*, 82 pp., Notes on Pure Mathematics No. 9, Dept. of Pure Mathematics and Dept. of Mathematics, I.A.S. Australian National University. Canberra. I.6.F. II.11.

HIGMAN, G. (See BOONE, W. W.)

HILBERT, D., 1890, Über die Theorie der algebraischen Formen, *Math. Annalen* **36**, 473–534. (*David Hilbert, Gesammelte Abhandlungen*, Vol. 2, 199–257, Julius Springer, Berlin, 1933.) II.8.

HILBERT, D., 1897, Die Theorie der algebraischen Zahlkörper, *Jahresbericht der deutschen Mathematiker-Vereinigung* **6**. Reprinted in: *David Hilbert. Gesammelte Abhandlungen*, Vol. 1, Julius Springer, Berlin, 1932. I.8. II.8.

HILBERT, D., 1900, See: Mathematical developments arising from Hilbert's problems, *Proc. of Symposium in Pure Mathematics* **28**, ed. F. Browder, American Math. Soc., Providence, RI 1978. I.4, 7.

HILBERT, D., 1901 and 1930, *Grundlagen der Geometrie*, Teubner, Leipzig, Berlin. The seventh edition (1930) has 326 pp. and contains much more material than the earlier editions. II.6.

HILBERT, D., 1932, *David Hilbert. Gesammelte Abhandlungen*, Vol. 1, Zahlentheorie, 539 pp, Julius Springer, Berlin. II.6.

HIRSCH, K. A., 1938, On infinite soluble groups I, *Proc. London Math. Soc.* (2) **44**, 53–60. I.6.F. II.6, 10.

HIRSCH, K. A., 1938 to 1954, On infinite soluble groups, *Proc. London Math. Soc.* (2) **44**, 53–60, 336–344; **49**, 184–194. *J. London Math. Soc.* **27**, 81–85; **29**, 250–251. II.6, 10.

HOARE, A. H. M., A. KARRASS, and D. SOLITAR, 1971 and 1973, Subgroups of finite index in Fuchsian groups, *Math. Z.* **120**, 289–298. Subgroups of infinite index in Fuchsian groups, *Math. Z.* **125**, 59–69. Subgroups of NEC-groups, *Comm. Pure Appl. Math.* **26**, 731–744. II.10.

HOARE, A. H. M., 1979, Coinitial graphs and Whitehead automorphisms, *Candian J. Math.* **31**, 112–123. II.2.

HÖLDER, O., 1895, Bildung zusammengesetzter Gruppen, *Math. Annalen* **46**, 321–422. II.9.

HOPF, H., 1930, Zur Algebra der Abbildungen von Mannigfaltigkeiten, *J. reine u. angew. Math.* **163**, 71–88. II.2.

HOPF, H., 1931, Beiträge zur Theorie der Flächenabbildungen, *J. reine u. angew. Math.* **165**, 225–236. II.2.

HOPF, H., 1942, Fundamentalgruppe und zweite Bettische Gruppe, *Comm. Math. Helvetici* **14**, 257–309. II.9.

HOPF, H., 1944, Enden offener Räume und unendliche diskontinuierliche Gruppen, *Comment. Math. Helvetici* **16**, 81–100. II.9.

HOPF, H., 1945, Über die Bettischen Gruppen, die zu einer beliebigen Gruppe gehören, *Comment. Math. Helvetici* **17**, 39–79. II.9.

HOROWITZ, R., 1972 and 1975, Characters of free groups represented in the two dimensional linear group, *Comm. Pure and Appl. Math.* **25**, 635–649 (1972). Induced automorphisms on Fricke characters of free groups, *Trans. American Math. Soc.* **208**, 41–55 (1975). I.6.E.

HOWE, R., 1930, On the role of the Heisenberg group in harmonic analysis, *Bull.* (*New Ser.*) *American Math. Soc.* **3**, 821–843. II.10.

HOYER, P, 1902, Über Definition und Behandlung transitiver Gruppen, *J. f. d. reine u. angew. Math.* **124**, 102–114. I.6.F.

HUA, L. K. and I. REINER, 1949, On the generators of the symplectic modular group, *Trans. American Math. Soc.* **65**, 415–426. I.6.A.

HUMBERT, M. G., 1915–1920. (a) Sur la réduction des formes d'Hermite dans un corps quadratique imaginaire. (b) Sur la formation du domaine fondamental d'un groupe automorphe. (c) Sur les réprésentations propres d'un entier par les formes positives d'Hermite dans un corps quadratique imaginaire. (d) Sur la mesure de l'ensemble des classes positives d'Hermite, de discriminant donné dans un corps quadratique imaginaire. (e) Sur la mesure des classes d'Hermite de discriminant donné dans un corps quadratique imaginaire, et sur certains volumes non euclidiens. (f) Expression de l'aire non euclidienne du domaine fondamentale lié à une forme d'Hermite indéfinie. All papers appeared in: *Comptes Rendues hebdomadaires de l'Académie des Sciences*, Paris, in the following volumes: (a) (1915) **161**, No. 8, 189–196, 227–234. (b) (1919) **168**, No. 5, 205–211. (c) (1919) **169**, 309–315. (d) (1919) **169**, 407–414. (e) (1919) No. 8, 448–454. (f) (1920) **171**, 377–382, 445–450. I.6.B.

HUMPHREYS, J. E., 1972, *Introduction to Lie Algebras and Representation Theory*, 169 pp., Graduate Texts in Mathematics **9**, Springer-Verlag, New York-Heidelberg-Berlin. I.6.E. II.7.

HUMPHREYS, J. E., 1975, *Linear Algebraic Groups*, 247 pp., Graduate Texts in Mathematics **21**, Springer-Verlag, New York-Heidelberg-Berlin. I.6.E.

HUREWITZ, W., 1931, Zu einer Arbeit von O. Schreier, *Abh. Math. Sem. Hamburg. Univ.* **8**, 307–314. II.3.

HURRELBRINK, J. and U. REHMANN, 1975, Eine endliche Präsentation der Gruppe $G_2(\mathbb{Z})$, *Math. Z.* **141**, 243–251. I.6.A.

HURWITZ, A., 1891, Über Riemannsche Flächen mit gegebenen Verzweigungspunkten (Part II, pp. 23–34), *Math. Annalen* **39**, 1–61. Reprinted in: *Mathematische Werke von Adolf Hurwitz*, Vol. 1., Birkhäuser Verlag, Basel, 1932. I.6.D.

HURWITZ, A., 1895, Die unimodularen Substitutionen in einem algebraischen Zahlkörper, *Nachrichten d. K. Gesellschaft d. Wissensch. zu Göttingen, Math. Phys. Kl*, 1895, 332–356. Reprinted in Vol. 1, *Mathematische Werke von Adolf Hurwitz*, Birkhäuser Verlag, Basel, 1932. I.6.A, D.

Hurwitz, A., 1905, Zur Theorie der automorphen Funktionen von beliebig vielen Veränderlichen, *Math. Annalen* **61**, 325–368. Reprinted in *Mathematische Werke von Adolf Hurwitz*, Vol. 1, Birkhäuser Verlag, Basel, 1932. I.6.A.

Ito, N., 1950, Note on *p*-groups, *Nagoya Math. J.* **1**, 113–116. II.6.
Iyanaga, S., 1934, Zum Beweis des Hauptidealsatzes, *Abh. Math. Sem. Hamburg. Univ.* **10**, 349–357. II.6.
Jacobson, N., 1937, Abstract derivation and Lie algebras, *Trans. American Math. Soc.* **42**, 206–224. II.7.
Jacobson, N., 1961, *Lie Algebras*, 331 pp. Interscience-Wiley, New York-London. II.7.
Jónsson, B. (See Federer, H.)
Jordan, C., 1870, Traité des substitutiones et des équations algébriques, Paris. I.6.F, 7, 8.
Jordan, C., 1874, Sur une application de la théorie des substitutions aux équations différentielles linéaires, *Comptes Rendues de l'Académie de France* **78**, 741–743. II.7.

Kaloujnine, L. and M. Krasner, 1948 and 1951, Le produit complet des groupes de permutations et le problème d'extension des groupes, *Comptes Rendues Acad. Sci. Paris* **227**, 806-808 (1948). Produit complet des groupes de permutations et le problème d'extensions des groupes III, *Acta Sci. Math. Szeged* **14**, 69–82 (1951). I.6.F. II.8.
Kaplansky, I., 1948, Rings with a polynomial identity, *Bull. American Math. Soc.* **54**, 575–580. II.8.
Karrass, A., W. Magnus, and D. Solitar, 1960, Elements of finite order in groups with a single defining relation, *Comm. Pure and Appl. Math.* **13**, 57–66. II.5.
Karrass, A. (See Hoare, A. H. M., Magnus, W.)
Kerber, A., 1971, Representation of permutation groups I, 192 pp, *Lecture Notes in Mathematics*, Vol. **240**, Springer-Verlag, Berlin-Heidelberg-New York. I.6.F.
Kervaire, M., 1965, Les neuds de dimensions supérieurs, *Bull. Soc. Math. France* **93**, 225–271. I.6.A. II.10.
Kleene, S. C., 1981, The theory of recursive functions approaching its centennial, *Bull. (New Ser.) Amer. Math. Soc.* **5**, 43–61. II.11.
Klein, F., 1872a, *Vergleichende Betrachtungen über neuere geometrische Forschungen*, 48 pp. Erlangen. Reprinted in *Felix Klein, Gesammelte Mathematische Abhandlungen*, Vol. 1, 460–497, Julius Springer, Berlin, 1921. Referred to in the literature as "Erlanger Programm." I.7.
Klein, F., 1872b, Über Liniengeometrie und metrische Geometrie, *Math. Annalen* **5**, 257–277. I.7.
Klein, F., 1880, Zur Theorie der elliptischen Modulfunctionen, *Math. Annalen* **17**, 62–70. I.6.B.
Klein, F., 1883, Neue Beiträge zur Riemanschen Functionentheorie, *Math. Annalen* **21**, 141–218 (esp. pp. 200–202). I.3, 7. II.4.
Klein, F. and R. Fricke, 1890 and 1892, *Vorlesungen über die Theorie der elliptischen Modulfunctionen*, 2 Volumes, Teubner. Leipzig. Reprinted: Johnson Corporation (Academic Press), 1966, New York. I.6.B.
Klein, F., 1926, *Vorlesungen über höhere Geometrie*, 3rd Edition, 405 pp., ed. W. Blaschke, Grundlehren der math. Wissenschaften 22, Springer-Verlag, Berlin. II.4.
Klein. F. (See Fricke, R.)
Koch, H., 1969, Zum Satz von Golod-Schafarewitsch, *Math. Nachr.* **42**, 321–333. II.6.
Kostrikin, A. I., 1947, On the connection between periodic groups and Lie rings, *Izv. Akad. Nauk. SSSR. Ser. Math.* **21**, 289–310 (Russian). *American Math. Soc. Translations (2)*, **45**, 165–189 (1965). II.7.
Kostrikin, A. I., 1958, The Burnside problem, *Izv. Akad. Nauk. SSSR. Ser. Math.* **23**, 3–34 (Russian). *American Math. Soc. Translations (2)* **36**, 63–99 (1964). II.7.
Kovács, L. G. and M. F. Newman, 1973, Hanna Neumann's problems on varieties of groups, Proc. 2nd Intern. Conference on the Theory of Groups, *Lecture Notes in Mathematics* 372, 417–431, ed. M. F. Newman, 740 pp., Springer-Verlag, Berlin-Heidelberg-New York.
Kra, I. (See Bers, L.)
Krull, W., 1928, Galoische Theorie der unendlichen algebraischen Erweiterungen, *Math. Annalen.* **100**, 687–698. II.9.

KUROSCH, A. (A. G. KUROSH), 1932, Zur Zerlegung unendlicher Gruppen, *Math. Ann.* **106**, 107–113. II.4.

KUROSCH, A., 1933, Über freie Produkte von Gruppen, *Math. Ann.* **108**, 26–36. II.4.

KUROSCH, A., 1934, Die Untergruppen der freien Produkte von beliebigen Gruppen, *Math. Annalen* **109**, 647–660. I.6.B, F. II.4.

KUROSCH, A., 1937, Zum Zerlegungsproblem der Theorie der freien Produkte, *Recueil Mathématique (Mat. Sbornik)* **2** (44): 5, 995–1001. II.4.

KUROSCH, A., 1939, Lokal freie Gruppen, *C. R. (Doklady) Acad. Sci. USSR. (N.S.)* **24**, 99–101. II.10.

KUROŠ, A. G. (KUROSH), 1944, *Teoriya Grupp* (Theory of Groups), OGIZ, Moscow-Leningrad, 372 pp (Russian). I.2. II.4, 9.

KUROŠ, A. G. 1955 and 1956, *The Theory of Groups*, Vol. 1, 272 pp., Vol. 2, 308 pp., translated from the Russian By K. A. Hirsch, Chelsea, New York. II.4.

KUROSH, A. G., 1963, *Lectures on General Algebra*, 335 pp., translated from the Russian by K. A. Hirsch, Chelsea, New York. II. 4.

KUROSH, A. (See DIETZMAN, A. P.)

LANDAU, E., 1927, *Vorlesungen über Zahlentheorie*, 3 volumes, S. Hirzel, Leipzig. II.14.

LAZARD, M., 1954, Sur les groupes nilpotents et les anneaux de Lie, *Annales Sci. L'Ecole Normale Supérieur.* (*3*) **71**, 101–190. II.7.

LERON, U., 1976, Trace identities and polynomial identities on $n \times n$ matrices, *J. Algebra* **42**, 369–377. I.6.E.

LEVI, F. W., 1929, *Geometrische Konfigurationen*, 310 pp., S. Hirzel, Leipzig. II.4.

LEVI, F. (LEVI, F. W.) 1930, Über die Untergruppen der freien Gruppen, *Math. Z.* **32**, 315–318. II.2, 3.

LEVI, F. and B. L. VAN DER WAERDEN, 1933, Über eine besondere Klasse von Gruppen, *Abh. Math. Sem. Hamburg. Univ.* **9**, 154–158. I.6.F.

LEVI, F., 1933, Über die Untergruppen der freien Gruppen II, *Math. Z.* **37**, 90–97. II.7.

LEVI, F. W., 1941, On the number of generators of a free product and a lemma of Alexander Kurosch, *J. Indian Math. Soc. (N.S.)* **5**, 149–155. II.4.

LEVI, F. W., 1942a, Ordered groups, *Proc. Indian Math. Akad. Sci.* **16**, 256–263. II.6.

LEVI, F. W., 1942b, Groups in which the commutator operation satisfies certain algebraic conditions, *J. Indian Math. Soc. (N.S.)* **6**, 87–97. II.7.

LEVI, F. W., 1944, Notes on group theory, IV–VI, *J. Indian Math. Soc. (N.S.)* **8**, 78–91. II.4.

LEVI, F. (See BAER, R.)

LEVIN, F., 1963, Factor groups of the modular group, *J. London Math. Soc.* **43**, 195–203. I.6.B.

LICKORISH, W. B. R., 1964. A finite set of generators for the homeotopy group of a 2-manifold, *Proc. Cambridge Philos. Soc.* **60**, 769–778. II.10.

LIE, S., 1872, Über Complexe, insbesondere Linien-und Kugelcomplexe, mit Anwendungen auf die Theorie der partiellen Differentialgleichungen, *Math. Annalen* **5**, 145–256. I.7.

LIE. S., 1888, *Theorie der Transformationsgruppen*, Vol. 1, 632 pp., Teubner, Leipzig. I.7. II.7.

LISTING, J. B., 1848, *Vorstudien zur Topologie*, Goettingen. I.4.

LOEWY, A., 1908, Die Rationalitätsgruppe einer linearen homogenen Differentialgleichung, *Math. Annalen* **65**, 129–160. I.6.E.

LOEWY, A., 1910, Algebraische Gruppentheorie, pp. 168–249 in *E. Pascal, Repertorium der höheren Mathematik*, Vol. 1, Analysis, ed. Paul Epstein, 2nd Ed, First part, Teubner, Leipzig, Berlin. I.2, 8.

LYNDON, R., 1950, Cohomology theory of groups with a single defining relation, *Annals. Math.* (*2*) **52**, 650–665. II.5.

LYNDON, R. C. and J. L. ULLMAN, 1969, Groups generated by two parabolic linear fractional transformations, *Canadian J. Math.* **21**, 1388–1403. I.3.

LYNDON, R. C. and P. E. SCHUPP, 1976, *Combinatorial Group Theory*, 339 pp., Ergebnisse der Mathematik **89**, Springer-Verlag, Berlin-Heidelberg-New York. I.4. II.2, 3, 4, 5, 6, 9, 10, 11, 14.

LYNDON, R. C. (See HIGGINS, R. J.)

MACBEATH, A. M., 1964, Groups of homeomorphisms of a simply connected space, *Annals Math.* (2) **79**, 473–488. I.6.B.

MACINTYRE, A., 1972, On algebraically closed groups, *Annals of Math.* **96**, 53–97. II.11.

MACLANE, S., 1963, *Homology*, 422 pp. (footnote, p. 137), Springer-Verlag, Berlin-Göttingen-Heidelberg, I.6.B. II.9.

MAGNUS, W., 1930, Über discontinuierliche Gruppen mit einer definierenden Relation (Der Freiheitssatz), *J. f. d. reine u. angew. Math.* **163**, 141–165. II.5.

MAGNUS, W., 1931, Untersuchungen über einige unendliche diskontinuierliche Gruppen, *Math. Annalen.* **105**, 52–74. I.5. II.5.

MAGNUS, W., 1932, Das Identitätsproblem für Gruppen mit einer definierenden Relation, *Math. Annalen* **106**, 295–307. II.5.

MAGNUS, W., 1934a, Über n-dimensionale Gittertransformationen, *Acta Math.* **64**, 353–367. I.6.A. II.2, 14.

MAGNUS, W., 1934b, Über den Beweis des Hauptidealsatzes, *J. reine u. angew. Math.* **170**, 235–240. II.6.

MAGNUS, W., 1934c, Über Automorphismen von Fundamentalgruppen berandeter Flächen, *Math. Annalen* **109**, 617–646. II.10.

MAGNUS, W., 1935, Beziehungen zwischen Gruppen und Idealen in einem speziellen Ring, *Math. Annalen* **111**, 259–280. II.2, 5, 6, 7.

MAGNUS, W., 1937a, Neuere Ergebnisse über auflösbare Gruppen, *Jahresber. Deutsche Math. Ver.* **47**, 69–78. II.7.

MAGNUS, W., 1937b, Über Beziehungen zwischen höheren Kommutatoren, *J. f. d. reine u. angew. Math.* **177**, 105–115. II.7.

MAGNUS, W., 1939a, Allgemeine Gruppentheorie, *Enzyklopädie der mathematischen Wissenschaften* I, 2nd edition, Issue 4, I, 51 pp., Teubner, Leipzig. I.11. II.14.

MAGNUS, W., 1939b, On a theorem of Marshall Hall, *Annals of Math.* **40**, 764–768. II.6.

MAGNUS, W., 1939c, Über freie Faktorgruppen und freie Untergruppen gegebener Gruppen, *Monatshefte f. Math. u. Physik* **47**, 307–313. II.5, 7.

MAGNUS, W., 1940, Über Gruppen und zugeordnete Liesche Ringe, *J. f. d. reine u. angew. Math.* **182**, 142–149. II.7.

MAGNUS, W., 1950, A connection between the Baker–Hausdorff formula and a problem of Burnside, *Annals of Math.* **52**, 111–126; **57**, 606 (1953). I.6. E, F. II.7.

MAGNUS, W. and RUTH MOUFANG, 1954, Max Dehn zum Gedächtnis, *Math. Annalen* **127**, 215–227. I.9.

MAGNUS, W., A. KARRASS, and D. SOLITAR, 1966, *Combinatorial Group Theory*, 444 pp, Wiley, New York, 2nd Ed. 1976. Dover, New York. I.2, 6.E, F. II.3, 4, 7.

MAGNUS, W., 1973, Rational representations of Fuchsian groups and nonparabolic subgroups of the modular group, *Nachr. Akad. Wiss. Göttingen II, Math. Phys. Kl*, **1973** No. 9, 179–189. II.3.

MAGNUS, W., 1974a, *Noneuclidean Tessellations and Their Groups*, 207 pp., Academic Press, New York. I.3, 6.B, C. II.3, 4, 10.

MAGNUS, W., 1974b, Braid Groups: A Survey. Proc. 2nd Internat. Conference on the Theory of Groups. *Lecture Notes in Mathematics* 372, pp. 463–487, Springer-Verlag, Berlin-Heidelberg-New York. I.6.D. II.10.

MAGNUS, W., 1975, Two-generator subgroups of PSL $(2, \mathbb{C})$, *Nachr. Akad. Wiss. Göttingen II, Math. Phys. Kl*, **1975**, No. 7, 81–94. II.5.

MAGNUS, W., 1978, Max Dehn, *The Mathematical Intelligencer* **1**, 132–142. I.9.

MAGNUS, W., 1980, Rings of Fricke characters and automorphism groups of free groups, *Math. Z.* **170**, 91–103. II.2.

MAGNUS, W. and CAROL TRETKOFF, 1980, Representations of automorphism groups of free groups, *Word Problems* II, ed. S. I. Adian, W. W. Boone, and G. Higman, 255–259. North Holland, Amsterdam-New York-Oxford. II.2.

MAGNUS, W., 1981, The uses of 2 by 2 matrices in combinatorial group theory. A survey, *Resultate der Mathematik* **4**, 171–192.

MAGNUS, W. (See KARRASS, A.)

MAL'CEV, A. I., 1940, On faithful representations of infinite groups of matrices, *Mat. Sb.* **8**, 405–422 (Russian). *American Math. Soc. Translations* (2) **45**, 1–18 (1965). I.6.E. II.5.

MAL'CEV, A. I., 1949, Nilpotent torsion free groups, *Izvestiya Akad. Nauk SSSR. Ser. Mat.* **13**, 201–212 (Russian). II.7, 10.

MAL'CEV, A. I., 1951, On certain classes of infinite soluble groups, *Mat. Sbornik* **28**, 567–588 (Russian). *American Math. Soc. Translations* (2) **2**, 1–21 (1956). II.6.

MARKOFF, A., 1945, Foundations of the algebraic theory of tresses, *Trav. Inst. Math. Stekloff* **16**, 53 pp. (Russian; English summary). II.10.

MASCHKE, H., 1896, The representation of finite groups, especially of the rotation groups of the regular bodies in three- and four-dimensional space by Cayley's color diagrams, *American J. Math.* **18**, 156–194. I.5.

MASKIT, B., 1965, On Klein's combination theorem, *Trans. American Math. Soc.* **120**, 499–509. I.3.

MASKIT, B., 1971, On Klein's combination theorem III, *Annals of Math. Studies* **66**, 297–316, Princeton University Press, Princeton, NJ. I.3.

MASSEY, W. S., 1967, *Algebraic Topology: An Introduction* (pp. 111ff), Harcourt, Brace and World, New York. II.4.

MATIYASEVICH, YU. V., 1971, Diophantine enumeration of enumerable predicates, *Izv. Akad. Nauk SSSR. Ser. Mat.* **35**, 3–30. II.11.

MCCARTHY, D., 1970, Infinite groups whose proper quotient groups are finite, *Comm. Pure and Appl. Math.* **23**, 767–790. I.6.A.

MCCOOL, J., 1974, A presentation of the automorphism group of a free group of finite rank, *J. London Math. Soc.* **8**, 259–266. II.2, 14.

MCCOOL, J. 1975a, On Nielsen's presentation of the automorphism group of a free group, *J. London Math. Soc.* **10**, 265–270. II.2, 14.

MCCOOL, J., 1975b, Some finitely presented subgroups of the automorphism group of a free group, *J. Algebra* **35**, 205–213. II.2, 14.

MEIER-WUNDERLI, H., 1950, Über endliche p-Gruppen, deren Elemente der Gleichung $x^p = 1$ genügen, *Comm. Math. Helvetici* **24**, 18–45. I.6.F.

MENDELSOHN, N. S. (See REE, R.)

MENNICKE, J., 1965, Finite factor groups of the unimodular group, *Annals of Math.* **31**, 31–37. I.6.B II.2.

MENNICKE, J. L. (MENNICKE, J.), Editor, 1980, Burnside groups, *Lecture Notes in Mathematics* **806** (Proceedings Bielefeld, 1977), 274 pp., Springer-Verlag, Berlin-Heidelberg-New York. II.7.

MIHAILOVA, K. A., 1958, The occurrence problem for direct products of groups, *Dokl. Akad. Nauk SSSR.* **119**, 1103–1105 (Russian). II.5.

MILLER, G. A., 1900, Report on the groups of infinite order, *Bull. American Math. Soc.* **7**, 121–130. Reprinted in: *Collected Works of George Abram Miller*, No. 64, Vol. 2. pp. 19–26. University of Illinois, Urbana, IL. I.8.

MILLER, G. A., 1935, History of the theory of groups to 1900, *The Collected Works of George Abram Miller*, Vol. 1, No. 62, pp. 427–467, University of Illinois, Urbana, IL. I.2, 8.

MILLER, G. A., 1938, Note on the history of group theory during the period covered by this volume, *The Collected Works of George Abram Miller*, Vol. 2, No. 63, pp. 1–18, University of Illinois, Urbana, IL. I.8.

MILLER, G. A., 1946–1955, *The Collected Works of George Abram Miller*, Vols. 3 and 4, The University of Illinois, Urbana, IL. I.8.

MILNOR, J. (See BASS, H.)

MINKOWSKI, H., 1887, Zur Theorie der positiven quadratischen Formen, *J. f. d. reine u. angew. Math.* **101**, 196–202. Reprinted in: *Gesammelte Abhandlungen von Hermann Minkowski*, Chelsea, New York, 1967. I.6.A. II.2.

MINKOWSKI, H., 1905, Diskontinuitätsbereich für arithmetische Äquivalenz, *J. reine u. angew. Math.* **129**, 220–274. Reprinted in *Gesammelte Abhandlungen von Hermann Minkowski*, Chelsea, New York, 1967. I.6.A., 8. II.2.

MINKOWSKI, H., 1973, *Hermann Minkowski, Briefe an David Hilbert*, eds. L. Ruedenberg and H. Zassenhaus, Springer-Verlag, Berlin-Heidelberg-New York. I.8.

MOUFANG, R., 1933, Alternativkörper und der Satz vom vollständigen Vierseit (D_9), *Abh. Math. Sem. Hamburg. Univ.* **9**, 207–222. II.6.

MOUFANG, R., 1934, Zur Struktur von Alternativkörpern, *Math. Annalen* **110**, 416–430. II.6.

MOUFANG, R., 1937, Einige Untersuchungen über geordnete Schiefkörper, *J. reine u. angew. Math.* **176**, 203–223. II.6.
MOUFANG, R. (See MAGNUS, W.)

NEUMANN, B. H., 1932, Die Automorphismengruppen der freien Gruppen, *Math. Annalen* **107**, 367–386. See also *Zentralblatt f. Math.* **5**, 244 (1933). II.14.
NEUMANN, B. H., 1933, Über ein gruppentheoretisch arithmetisches Problem, *S.-B. Preussische Akademie d. Wiss. Phys.-Math. Kl*, 18 pp. (For a survey of results, see MAGNUS [1974(a)]. I.6.B. II.3.
NEUMANN. B. H., 1937a, Identical relations in groups I, *Math. Annalen* **114**, 506–525. II.8.
NEUMANN, B. H., 1937b, Some remarks on infinite groups, *J. London Math. Soc.* **12**, 120–127. I.7. II.4, 11.
NEUMANN, B. H., 1943a, Adjunction of elements to groups, *J. London Math. Soc.* **18**, 4–11. II.5, 11.
NEUMANN, B. H., 1943b, On the number of generators of a free product, *J. London Math. Soc.* **18**, 12–20. II.4.
NEUMANN, B. H., 1949a, On ordered groups, *American J. Math.* **71**, 1–18. II.6.
NEUMANN, B. H., 1949b, On ordered division rings, *Trans. American Math. Soc.* **66**, 202–252. II.6.
NEUMANN, B. H., 1950, A two-generator group isomorphic to a proper factor group, *J. London Math. Soc.* **25**, 247–248. II.5.
NEUMANN, B. H., 1954, An essay on free products with amalgamations, *Phil. Trans. Royal Soc. London* **246**, No. 919, 503–554 (June 15, 1954). II.4.
NEUMANN, B. H., 1956, Ascending derived series, *Compositio Math.* **13**, 47–64. II.8.
NEUMANN, B. H., 1973, The isomorphism problem for algebraically closed groups, pp. 553–562 in *Word Problems*, North Holland, Amsterdam. II.11.
NEUMANN, B. H. (See also HIGMAN, G.)
NEUMANN, H., 1967, *Varieties of Groups*, 192 pp. Ergebnisse der Mathematik und ihrer Grenzgebiete 37, Springer-Verlag, Berlin-Heidelberg-New York. I.6.F. II.8.
NEUMANN, H. (See also HIGMAN, G.)
NEWMAN, B. B., 1968, Some results on one-relator groups, *Bull. American Math. Soc.* **74**, 568–571. II.5.
NEWMAN, M. F. (See KOVASC, G.)
NEUWIRTH, L. P., 1965, *Knot Groups*, 111 pp. Annals of Mathematics Studies 56, Princeton University Press, Princeton, NJ. II.10.
NIELSEN, J., 1913, Kurvennetze auf Flächen, 58 pp., Ph.D. Thesis, University of Kiel (Germany), privately printed. I.6.D, 7. II.2, 10.
NIELSEN, J., 1917, Die Isomorphismengruppe der allgemeinen unendlichen Gruppe mit zwei Erzeugenden, *Math. Annalen* **78**, 385–397. II.2.
NIELSEN, J., 1918, Über die Isomorphismen unendlicher Gruppen ohne Relation, *Math. Annalen* **79**, 269–272. II.2.
NIELSEN, J., 1921, Om Regning med ikke kommutative Faktorer og dens Anvendelse i Gruppeteorien, *Matematisk Tidsskrift* **B**, 77–94. English translation: *Math. Scientist* **6**, 73–85 (1981). I.6.D. II.2, 5, 7.
NIELSEN, J., 1924a, Die Isomorphismengruppe der freien Gruppen, *Math. Annalen* **91**, 169–209. I.6.A. II.2, 14.
NIELSEN, J., 1924b, Die Gruppe der dreidimensionalen Gittertransformationen, *Det. Kgl. Danske Videnskabernes Selskab. Mat.-Fys. Meddelelser.* **5**, 12, 1–29. I.6.A. II.2.
NIELSEN, J., 1927–1931, Untersuchungen zur Topologie der geschlossenen zweiseitigen Flächen. I, II, III, *Acta Math.* **50**, 189–358; **53**, 1–76; **58**, 87–167. II.10.
NIELSEN, J., 1955, A basis for subgroups of free groups, *Math. Scand.* **3**, 31–43. II.2.
NOETHER, E., 1918, Invariante Variationsprobleme, *Nachr. K. Ges. d. Wiss. zu Göttingen. Math. Phys. Kl*, 1–23. I.6.E.
NOETHER, E., 1921, Idealtheorie in Ringbereichen, *Math. Annalen* **83**, 24–66. II.6.
NOVIKOV, P. S. (NOVIKOFF), 1955, On the algorithmic unsolvability of the word problem in group theory (Russian), *Trudy Mat. Inst. im. Steklov* No. 44., Izdat Akad. Nauk. SSSR.,

Moscow, 143 pp. English translation in: *American Math. Soc. Translations* **9**, 122 pp. (1958). II.2, 11.

OATES, SHEILA and M. B. POWELL, 1964, Identical relations in finite groups, *J. Algebra* **1**, 11–39. II.8.

OL'ŠANSKIĬ, A. JU., 1970, On the problem of a finite basis for the identities of groups, *Izv. Akad. Nauk. SSSR.* **34**, 376–384 (Russian). English Transl. in: *Math. USSR-Izv.* **4**, 381–389 (1971). II.8.

PAPAKYRIAKOPOULOS, CH. D., 1957, On Dehn's lemma and the asphericity of knots, *Annals of Math.* (2) **66**, 1–26. I.4.

PEIFFER, RENÉE, 1949, Über Identitäten zwischen Relationen, *Math. Annalen* **121**, 67–99. II.5.

PICARD, E., 1896, *Traité d'Analyse*, Vol. 3., XIV + 568 pp., Gauthier-Villars, Paris (pp. 531 ff). I.6.B, E.

PICK, G., 1886, Über gewisse ganzzahlige lineare Substitutionen welche sich nicht durch algebraische Congruenzen erklären lassen, *Math. Annalen* **28**, 119–124. I.6.B.

PICKEL, P. F., 1971, Finitely generated nilpotent groups with isomorphic finite quotients, *Trans. American Math. Soc.* **160**, 327–341. I.6.E.

POINCARÉ, H., 1882, Théorie des groupes fuchsiens, *Acta Math.* **1**, 1–62. I.3, 6.E.

POINCARÉ, H., 1883, Mémoire sur les groupes kleinéens, *Acta Math.* **3**, 49–92. I.6.E.

POINCARÉ, H., 1884, Sur les groupes des équations linéaires, *Acta Math.* **4**, 201–312. I.6.E.

POINCARÉ, H., 1895, Analysis Situs (§12), *Journal d'Ecole Polytechnique Normale* **1**, 1–121. Reprinted in: *Oeuvres de Henri Poincaré*, 11 Volumes, Gauthier-Villars, Paris, 1928–1956. (Vol. 6, pp. 193–288, esp. pp. 239–242). I.4. II.9.

POINCARÉ, H., 1899, Sur les groupes continues, *Trans. Cambridge Philos. Soc.* **18**, 220–255. *Oeuvres de Henri Poincaré*, **3**, 173–212, Gauthier-Villars, Paris. II.7.

POINCARÉ, H., 1904, Cinquième complement a l'analysis situs, *Rendiconti de Circulo mathematico di Palermo* **18**, 45–110. *Oeuvres de Henri Poincaré*, Vol. **6**, 435–498, Gauthier-Villars, Paris. I.4.

POLYA, G., 1937, Kombinatorische Anzahlbestimmung für Gruppen, Graphen und chemische Verbindungen, *Acta Math.* **68**, 145–254 (p. 178). II.8.

PONTRIAGIN, L., 1934, The theory of topological commutative groups, *Annals of Math.* (2) **35**, 361–388. II.6, 9.

PONTRIJAGIN, L., 1939, *Topological Groups*, Oxford University Press, London. II.9.

POST, E. L., 1944, Recursively enumerable sets of positive integers and their decision problems, *Bull. American Math. Soc.* **50**, 284–316. II.11.

POWELL, M. B. (See OATES, S.)

PRIDE, S. J., 1977a and b (a) The two-generator subgroups of one-relator groups with torsion, *Trans. American Math. Soc.* **234**, 483–496. (b) The isomorphism problem for two-generator one-relator groups with torsion is solvable, *Trans. American Math. Soc.* **227**, 109–139. II.5.

PROCESI, C., 1976, The invariant theory of *n* by *n* matrices, *Advances in Math.* **19**, 306–381. I.6.E.

RADO, T. (See HALL, M.)

RAPAPORT, ELVIRA STRASSER, 1958, On free groups and their automorphisms, *Acta Math.* **99**, 139–163. II.2.

RAPAPORT, E. S., 1968, Groups of order 1. Some properties of presentations, *Acta Math.* **121**, 127–150. II.5.

RAZMYSLOV, JU. P., 1974, Identities with traces in full matrix algebras over a field of characteristic zero (Russian), *Izv. Akad. Nauk. SSSR. Ser. Math.* **38**, 723–756. I.6.E.

REE, R. and N. S. MENDELSOHN, 1974, Free subgroups of groups with a single defining relation, *Archiv d. Math.* **19**, 577–580. II.5.

REID, CONSTANCE, 1978, The road not taken, *The Mathematical Intelligencer* **1**, 21–23. I.8.

REIDEMEISTER, K., 1921, Relativklassenzahl gewisser relativ-quadratischer Zahlkörper, *Abh. Math. Sem. Hamburg. Univ.* **1**, 27–48. II.3.

REIDEMEISTER, K., 1926, Knoten und Gruppen, *Abh. Math. Sem. Hamburg. Univ.* **5**, 7–23. I.3, 6.E.

REIDEMEISTER, K., 1927, Über unendliche diskrete Gruppen, *Abh. Math. Sem. Hamburg. Univ.* **5**, 33–39. I.6.E. II.7.

REIDEMEISTER, K., 1932a, *Knotentheorie*. Ergebnisse der Mathematik 1, No. 1., 74 pp., Springer-Verlag, Berlin. I.4. II.3, 6, 10.

REIDEMEISTER, K., 1932b, *Einführung in die kombinatorische Topologie.*, 209 pp., F. R. Vieweg & Sohn, Braunschweig. II.3, 5, 9.

REIDEMEISTER, K. and H. G. SCHUMANN, 1934, *L*-Polynome von Verkettungen, *Abh. Math. Sem. Hamburg. Univ.* **10**, 256–262. II.6.

REINER, L. (See HUA, L. K.)

REMAK, R., 1911, Über die Zerlegung der endlichen Gruppen in direkt unzerlegbare Faktoren, *J. f. d. reine u. angew. Math.* **139**, 293–308. I.6.D.

RIEMANN, B., 1857, Theorie der Abelschen Functionen, *J. f. d. reine u. angew. Math.* **54**. Reprinted in RIEMANN [1892], pp. 88–144. II.9.

RIEMANN, B., 1892, *Gesammelte mathematische Werke*, eds. H. Weber and R. Dedekind. Reprint: Dover, New York, 1953 as: *The Collected Works of Bernhard Riemann*. (The text is the German original.) I.7.

RIPS, I. A., 1972, On the fourth integer dimension subgroup, *Israel J. Math.* **12**, 342–346. II.7.

RITT, J. F., 1950, *Differential Algebra*, American Math. Soc. Colloquium Publications **33**, American Math. Soc., New York. VIII plus 184 pp. I.6.E.

ROBINSON, DEREK J. S., 1972, *Finiteness Conditions and Generalized Soluble Groups*, 2 Vols., Ergebnisse der Mathematik 62, 63, Springer-Verlag, New York-Heidelberg-Berlin. I. 6.F. II.6.

ROTA, G.-C., 1964, Theory of Moebius functions, *Z. Wahrscheinlichkeitstheorie u. verw. Gebiete.* **2**, 340–386. II.3.

ROTHAUS, O. S. (See GERSTENHABER, M.)

ROTMAN, J. J., 1973, *The Theory of Groups*, 2nd Ed., 342 pp., Allyn & Bacon, Boston. II.10, 11.

ŠAFAREVIČ, I. R., 1963, *Extensions with Prescribed Branch Points*. (Russian, French summary). Inst. Hautes Etudes Sci. Publ. Math. No. 18, 71–95. II.6.

ŠAFAREVIČ, I. R. (SCHAFAREVITSCH), 1973, Über einige Tatsachen in der Entwicklung der Mathematik (Russian original and German translation), *Jahrbuch der Akademie der Wissenschaften zu Göttingen*, 1973, pp. 31–42. English translation: *The Math. Intelligencer* **3**, 182–184 (1981). Preface.

ŠAFAREVIČ, I. R. (See also GOLOD, E. S.)

SANDLING, R., 1972 a. The dimension subgroup problem, *J. Algebra* **21**, 216–231. b. Dimension subgroups over arbitrary coefficient rings. *J. Algebra* **21**, 250–261. c. Modular augmentation ideals. *Proc. Cambridge Philos. Soc.* **71**, 25–32. II.7.

SANOV, I. N., 1940, Solution of Burnside's problem for $n = 4$, *Leningrad State University Annals* (*Uchenyi Zapiski*) *Math. Ser.* **10** 166–170 (Russian). I.6.F.

SANOV, I. N., 1949, Application of Lie rings to the theory of periodic groups, *Uspehi Mat. Nauk.* (*N.S.*) **4**, No. 3 (31), 180 (Russian). II.7.

SANSONE, G., 1923, I sottogruppi de gruppo di Picard e due teoremi sui gruppi finiti analoghi al teorema del Dyck, *Rend. Circulo Mat. Palermo* **47**, 273–333. I.6.B.

SCHLESINGER, L., 1909, *Bericht über die Entwicklung der Theorie der linearen Differentialgleichungen seit* 1865, 133 pp., 1742 references. Ergänzungsband III zu dem 18. Bande des Jahresberichts der Deutschen Mathematiker Vereinigung, Teubner, Leipzig, Berlin. I.6.E. II.14.

SCHMID, W., 1982, Poincaré and Lie groups, *Bull.* (*New Ser.*) *American Math. Soc.* **6**, 175–186.

SCHMIDT, A., 1934, Die Herleitung der Spiegelung aus der ebenen Bewegung, *Math. Annalen* **109**, 538–571. II.3.

SCHMIDT, O. (SCHMIDT, O. J., SCHMIDT, O. JU.), 1913, Sur les produits directs, *Bull. de la Soc. Math. de France* **41**, 161–164. I.6.F.

SCHMIDT, O. J., 1916, *Abstract Group Theory* (Russian), 214 pp., Kiev. I.6.F. II.4.

SCHMIDT, O., 1928, Über unendliche Gruppen mit endlicher Kette, *Math. Z.* **29**, 34–41. II.4.

SCHMIDT, O. JU., 1933, *Abstract Theory of Groups*, 2nd Ed., U.S.S.R. State Technical and Theoretical Publishing House. English Translation, 174 pp., W. H. Freeman & Co., San Francisco, 1966. I.6.F.

SCHOENFLIES, A. and M. DEHN, 1930, *Einführung in die analytische Geometrie der Ebene und des Raumes*, 2nd Ed., 414 pp., Julius Springer, Berlin. II.3.

SCHOLZ, A. and OLGA TAUSSKY, 1934, Die Hauptideale der kubischen Klassenkörper imaginärquadratischer Zahlkörper, *J. reine u. angew. Math.* **171**, 19–41. II.6.

SCHREIER, O., 1924, Über die Gruppen $A^a B^b = 1$, *Abh. Math. Seminar Hamburg. Univ.* **3**, 167–169. II.3, 4.

SCHREIER, O., 1926 a and b, Über die Erweiterung von Gruppen I, II. *Monatshefte f. Math. u. Phys.* **34**, 165–180 and *Abh. Math. Seminar Hamburg. Univ.* **4**, 321–346. II.3, 6, 9.

SCHREIER, O., 1927a, Die Untergruppen der freien Gruppen, *Abh. Math. Sem. Hamburg Univ.* **5**, 161–183. I.2, 5, 6.F II.2, 3, 4, 5.

SCHREIER, O., 1927b, Die Verwandtschaft stetiger Gruppen im Grossen, *Abh. Math. Seminar Hamburg. Univ.* **5**, 233–244. II.3.

SCHREIER, O., 1928, Über den Jordan–Hölderschen Satz, *Abh. Math. Sem. Hamburg. Univ.* **6**, 300–302. II.3.

SCHREIER, OTTO (Obituary), 1930, In Vol. **7** of *Abh. Math. Seminar Hamburg. Univ.*, after page 106, anonymous. II.3.

SCHREIER, O. and E. SPERNER, 1931 and 1935, *Einführung in die analytische Geometrie und Algebra*, 2 Vols., Teubner, Leipzig, Berlin, Partly translated into English and published as *Introduction to Modern Algebra and Matrix Theory*, Chelsea, New York, 1951. II.3.

SCHUMANN, H. G., 1933, Zum Beweis des Hauptidealsatzes, *Abh. Math. Sem. Hamburg. Univ.* **12**, 42–47. II.6.

SCHUMANN, H. G., 1937, Über Moduln und Gruppenbilder, *Math. Ann.* **114**, 385–413. II.6.

SCHUMANN, H. G. (See also REIDEMEISTER, K.)

SCHUPP, P. E., 1971, Small cancellation theory over free products with amalgamations, *Math. Annalen* **193**, 255–264. II.4.

SCHUPP, P. E. (See also LYNDON, R. C.)

SCHUR, I., 1904, Über die Darstellung der endlichen Gruppen durch gebrochen lineare Substitionen, *J. reine u. angew. Math.* **127**, 20–50. Reprinted in *Issai Schur. Gesammelte Abhandlungen*, **1**, 86–116, Springer-Verlag, New York-Heidelberg-Berlin, 1973. I.6.F.

SCOTT, P. (See DYER, J. L.)

SCOTT, W. R., 1951, Algebraically closed groups, *Proc. American Math. Soc.* **2**, 118–121. II.5, 11.

SEIFERT, H., 1934, Über das Geschlecht von Knoten, *Math. Annalen* **110**, 571–592. II.6.

SEIFERT, H. and W. THRELFALL, 1934, *Lehrbuch der Topologie*, 353 pp., Teubner, Leipzig. Reprinted by Chelsea, New York, 1947. II.9, 10.

SEKI, TAKEJIRO, 1941, Über die Existenz der Zerfällungsgruppe in der Erweiterungstheorie der Gruppen, *Tohoku Math. J.* **48**, 235–238. II.9.

SELBERG, A., 1960, On discontinuous groups in higher dimensional symmetric spaces, Contributions to Function Theory, Internat. Colloquium Function Theory, Tata Institute of Fundamental Research, Bombay (pp. 147–164). I.6.E.

SERRE, J.-P., 1964, *Cohomologie galoisienne*, Lecture Notes in Mathematics 5, 212 pp., Springer-Verlag, Berlin-Heidelberg-New York. II.6.

SERRE, J.-P., 1977, Arbres, amalgames, SL_2, *Astérisque* **46** (Soc. Math. de France), 189 pp. II.4, 10.

SIEGEL, C. L., 1935, Über die analytische Theorie der quadratischen Formen, *Annals of Math.* **36**, 527–606. Reprinted in Vol. 1 of *Carl Ludwig Siegel. Gesammelte Abhandlungen*, Springer-Verlag, Berlin-Heidelberg-New York, 1966. I.6.A.

SIEGEL, C. L., 1939, Einführung in die Theorie der Modulfunktionen n-ten Grades, *Math. Annalen* **116**, 617–657. Reprinted in Vol. 2 of *Carl Ludwig Siegel. Gesammelte Abhandlungen*, Springer-Verlag, Berlin-Heidelberg-New York, 1966. I.6.A.

SIEGEL, C. L., 1943a, Symplectic geometry, *American J. Math.* **65**, 1–86. Reprinted as a book: Academic Press, New York, 1964. Reprinted also in Vol. 2, *Carl Ludwig Siegel. Gesammelte Abhandlungen*. Springer-Verlag, Berlin-Heidelberg-New York, 1966. I.6.A.

SIEGEL, C. L., 1943b, Discontinuous groups, *Annals of Math.* **44**, 674–689. Reprinted in Vol. 2 of *Carl Ludwig Siegel. Gesammelte Abhandlungen*, Springer-Verlag, Berlin-Heidelberg-New York, 1966. I.6.A.

SIEGEL, C. L., 1965, Zur Geschichte des Frankfurter Mathematischen Seminars, Victorio Kloosterman, Frankfurt am Main. Reprinted in Vol. 3 of *Carl Ludwig Siegel. Gesammelte Abhandlungen*. pp. 462–474, Springer-Verlag, Berlin-Heidelberg-New York, 1966. English translation: *Math. Intelligencer* **1**, 223–230 (1979). I.9.

ŠMELKIN, A. L., 1965a, On soluble products of groups (Theorem 3) (Russian), *Sibirsk. Math. Ž.* **6**, 212–220. II.6.

ŠMELKIN, A. L., 1965b, Wreath products and varieties of groups (Russian), *Izv. Akad. Nauk. SSSR. Ser. Mat.* **29**, 149–170. II.8.

SOLITAR, D. (See BAUMSLAG, G., HOARE, A. H. M., KARRASS, A., and MAGNUS, W.)

SPERNER, E. (See SCHREIER, O.)

STALLINGS, J. R., 1968, On torsion free groups with infinitely many ends, *Annals of Math.* (*2*) **88**, 312–334. I.5. II.9.

STAMMBACH, U., 1967, Ein neuer Beweis eines Satzen von Magnus, *Proc. Cambridge Philos. Soc.* **63**, 929–930. II.7.

STEINBERG, A., 1964, On free nilpotent quotient groups, *Math. Z.* **85**, 185–196. II.5.

STICKELBERGER, L. (See FROBENIUS, F. G.)

STILLWELL, J., 1980, *Classical Topology and Combinatorial Group Theory*, 301 pp., 305 figures, Graduate Texts in Mathematics **72**, Springer-Verlag, New York-Heidelberg-Berlin. II.9, 10.

STILLWELL, J., 1982, The word problem and the isomorphism problem for groups, *Bull.* (*New Ser.*) *American Math. Soc.* **6**, 33–56.

STORK, D., 1972, The action of the automorphism groups of F_2 upon the A_6 and $PSL(2,7)$-defining subgroups of F_2, *Trans. American Math. Soc.* **172**, 111–117. II.3.

STOUFF, X., 1888, Sur la transformation des functions fuchsiennes, *Annals de l'Ecole normale supérieure* (*3*) **5**, 219–326 (Thèse, Paris). II.2.

SWAN, R. G., 1967, Representations of polycyclic groups, *Proc. American Math. Soc.* **18**, 573–574. I.6.

SWAN, R. G., 1971, Generators and relations for certain special linear groups, *Advances in Math.* **6**, 1–77. I.6.B.

TARTAKOVSKII, V. A., 1949, The sieve method in group theory, *Mat. Sb.* **25** (67), 3–50 (Russian). English translation, together with that of related papers in *American Math. Soc. Transl.* **60**, 1952 (110 pp). II.9.

TAUSSKY, O., 1937, A remark on the class field tower, *J. London Math. Soc.* **12**, 82–85. II.6.

TAUSSKY, O. (See also SCHOLZ, A.)

THOMPSON, J. G. (See FEIT, W.)

THOMPSON, RICHARD J., 1980, Embeddings into finitely generated simple groups which preserve word problems, pp. 401–414 in: *Word Problems II.*, eds. S. I. Adian, W. W. Boone, and G. Higman, North Holland, Amsterdam. II.11

THUE, A., 1906–1914. a. Uber unendliche Zeichenreihen, Christiania Vidensk.-Selsk. Skr. 906, No. 7, 22 pp. (1906). b. Die Lösung eines Spezialfalles eines generellen logischen Problems, Kra Vidensk. Selsk. Skrifter I. Mat. Nat. Kl. No. 8. Kra. (1910). c. Probleme über Veränderungen von Zeichenreihen nach gegebenen Regeln, Kra. Vidensk. Selsk. Skrifter I. Mat. Nat. Kl. No. 10 (1914). These three papers have been reprinted in: *Selected Mathematical Papers of Axel Thue*, Universitetsforlaget Oslo, Bergen, Tromsø, 1977. They appear, respectively, as papers No. 8, 17, and 28. The last paper introduces what is now called "Thue systems." I.7. II.11.

TIETZE, H., 1908, Über die topologischen Invarianten mehrdimensionaler Mannigfaltigkeiten, *Monatshefte für Mathematik and Physik* **19**, 1–118. I.4. II.13.

TITS, J., 1972, Free subgroups in linear groups, *J. Algebra* **20**, 250–270. I.6.E. II.9.

TODD, J. A. and H. S. M. COXETER, 1936, A practical method for enumerating cosets of a finite abstract group, *Proc. Edinburg Math. Soc.* (*2*) **5**, 25–34. II.3.

TOLMIERI, R. (See AUSLANDER, L.)

TRETKOFF, C., 1975, Nonparabolic subgroups of the modular group, *Glasgow Math. J.* **16**, 90–102. II.3.

TRETKOFF, C. (See also MAGNUS, W.)

TURING, A. M., 1938, The extensions of a group, *Compos. Math.* **5**, 357–367. II.9.

ULLMAN, J. L. (See LYNDON, R. C.)

ULM, H., 1933, Zur Theorie der abzählbar unendlichen Abelschen Gruppen, *Math. Annalen* **107**, 774–803. II.6.

UTZKOW, A. T. (See DIETZMAN, A. P.)

VAN DER WAERDEN, B. L. (See LEVI, F.)

VAN KAMPEN, E. R., 1933a. On the connection between the fundamental groups of some related spaces, *American J. Math.* **55**, 261–267. b. On some lemmas in the theory of groups, *American J. Math.* **55**, 268–273. II.3, 4, 9, 13.

VAUGHAN-LEE, M. R., 1970, Uncountably many varieties of groups, *Bull. London Math. Soc.* **2**, 280–286. II.8.

VESSIOT, E., 1892, Sur l'integration des équations différentielles linéaires, *Ann. de l'Ecole Normale* (*3*) **9**, 197–280. (Thèse, Paris, Gauthier-Villars et Fils, 84 pp.) I.6.E.

VOGT, H., 1889, Sur les invariants fondamentaux des équations différentielles linéaires du second ordre, *Annales de l'Ecole Normale Supérieur* (*3*) **6**, Suppl. 3–72 (Thèse, Paris) I.6.E.

WAGNER, W., 1937, Über die Grundlagen der Geometrie und allgemeine Zahlsysteme, *Math. Annalen* **113**, 528–567. II.8.

WALDHAUSEN, W., 1968, The word problem in fundamental groups of sufficiently large irreducible 3-manifolds *Annals of Math.* **88**, 272–280. II.10.

WALDINGER, H., 1965, On the subgroups of the Picard group, *Proc. American Math. Soc.* **16**, 1373–1378. I.6.B.

WEBER, H. 1894 and 1896, *Lehrbuch der Algebra*, Vol. 1,2, Vieweg & Sohn, Braunschweig (appeared in 3 volumes). I.8. II.14.

WEBER, H., 1899, *Lehrbuch der Algebra*, 2 Vols, Vieweg & Sohn, Braunschweig. II.14.

WEHRFRITZ, B. A. F., 1973, *Infinite Linear Groups*, Ergebnisse der Mathematik 76, Springer-Verlag, New York-Heidelberg-Berlin. I.6.E. II.9.

WEISSNER, L., 1935, Abstract theory of inversion of finite series, *Trans. American Math. Soc.* **38**, 474–484. II.3.

WEVER, F., 1950, Über Regeln in Gruppen, *Math. Annalen* **122**, 334–339. II.7, 8.

WEYL, H., 1939, *The Classical Groups*, 302 pp., Princeton University Press and Oxford University Press. I.6.A.

WHITEHEAD, J. H. C., 1936a. On certain sets of elements in a free group, *Proc. London Math. Soc.* **41**, 48–56. b. On equivalent sets of elements in a free group, *Annals of Math.* **37**, 782–800. II.2.

WILSON, J. S., 1971, Groups with every proper quotient group finite, *Proc. Cambridge Philos. Soc.* **69**, 373–391. I.6.B.

WIRTINGER, W., 1905, Über die Verzweigungen bei Funktionen von zwei Veränderlichen, *Jahresbericht d. Deutschen Mathematiker Vereinigung.* **14**, 517. (Nothing but this title of a talk given by the author at a meeting is mentioned here.) I.4. II.3.

WITT, E., 1936, Bemerkungen zum Beweis der Hauptidealsatzes von S. Iyanaga, *Abh. Math. Seminar Hamburg. Univ.* **11**, 221. II.6.

WITT, E., 1937, Treue Darstellung Liescher Ringe, *J. reine u. angew. Math.* **177**, 152–160. II.7.

WUSSING, H., 1969, *Die Genesis des abstrakten Gruppenbegriffs*, 285 pp., V. E. B. Deutscher Verlag der Wissenschaften, Berlin. An English translation is in preparation and to be published by MIT-Press, Cambridge, MA. I.2, 11.

ZASSENHAUS, H., 1934, Zum Satz von Jordan–Hölder–Schreier, *Abh. Math. Seminar Hamburg. Univ.* **10**, 106–108. II.3.

ZASSENHAUS, H., 1937 and 1956, *Lehrbuch der Gruppentheorie*, Hamburger math. Einzelschriften, Teubner, Leipzig, Berlin. Expanded English translations appeared 1949 and 1956 under the title: *The Theory of Groups*, 265 pp., Chelsea, New York. I.2. II.6, 9.

ZASSENHAUS H., 1939, Über Liesche Ringe mit Primzahlcharakteristik, *Abh. Math. Seminar Hans. Univ. Hamburg* **13**, 1–100. II.7.

ZASSENHAUS, H., 1940, Ein Verfahren, jeder endlichen Gruppe einen Lie-Ring mit der Charakteristik p zuzuordnen, *Abh. Math. Seminar Hans. Univ. Hamburg* **13**, 200–207. II.7.

Index of Names*

---

*Does not include the bibliography or the biographical notes in Chapter I.9.

Index of Subjects